Robert E.C. Stearns

List of Papers Contributions to Natural History, etc.

chiefly on the conchology of North America

Robert E.C. Stearns

List of Papers Contributions to Natural History, etc.
chiefly on the conchology of North America

ISBN/EAN: 9783337426613

Printed in Europe, USA, Canada, Australia, Japan

Cover: Foto ©ninafisch / pixelio.de

More available books at **www.hansebooks.com**

CONTRIBUTIONS TO

NATURAL HISTORY, Etc.

CHIEFLY ON THE

CONCHOLOGY OF NORTH AMERICA.

By ROBERT E. C. STEARNS

PREFATORY NOTE.

The following will serve as an Index to papers of mine published from time to time, prior to January 1st, 1878, in the Proceedings of the California Academy and other Scientific Societies, or elsewhere as mentioned, with the date of publication.

INDEX.

1. List of Shells collected at Baulines Bay, California, June, 1866.

   Proceed. Cal. Acad. Nat. Sciences, July 16, 1866.

2. List of Shells collected at Santa Barbara and San Diego by Mr. J. Hepburn, in February-March, 1866, with remarks upon some of the species.

   Proceed. Cal. Acad. Nat. Sciences, August 6, 1866.
   [The above printed separately and distributed with the title of "Conchological Memoranda," 8 pp.]

3. Remarkable Instance of Vitality in a Snail.

   Proceed. Cal. Acad. Nat. Sciences, March 4, 1867.

4. Shells collected at Santa Barbara by W. Newcomb, M.D., in January, 1867.

   Proceed. Cal. Acad. Nat. Sciences, April 1, 1867.
   [The above also includes a collection made by Dr. Newcomb at Santa Cruz Island, Cal.]

5. List of Shells collected at Purissima and Lobitas, California, October, 1866.

   Proceed. Cal. Acad. Nat. Sciences, April 1, 1867.
   [Numbers 3, 4, 5 printed separately as "Conchological Memoranda, No. 2," which contains also brief notes on California Shells. 7 pp.]

6. List of Shells collected at Bodega Bay, California, June, 1867.

   Proceed. Cal. Acad. Nat. Sciences, December 2, 1867.

7. Shells collected by the U. S. Coast Survey Expedition to Alaska, in the year 1867.

Proceed. Cal. Acad. Nat. Sciences, December 2, 1867.
[Numbers 6 and 7 also printed separately as "Conchological Memoranda. No. III," 4 pp, 8vo. See also "Report of the Superintendent of the United States Coast Survey for 1867, Appendix No. 18," pp. 291-2, Harford's List, which contains numerous typographical errors not in my original.]

8. Shell Money.

American Naturalist, Vol. III, No. 1, March, 1869. [Contains notes on Western American Shells, used as money by the aborigines of California, etc. 5 pp.]

9. The Haliotis, or Pearly Ear-Shell.

American Naturalist, Vol. III, No. 5, July, 1869, with figures. [Includes notes on Californian species.]

10. Rambles in Florida.

American Naturalist, Vol. III, Nos. 6, 7, 8 and 9, 1869, 37 pp. [Containing many notes on the Natural History, Ethnology, etc., of Florida, made during a visit in January-March, 1869, in company with the late Dr. William Stimpson and Col. Ezekiel Jewett.]

11. On a new species of Pedipes from Tampa Bay, Florida.

Proceed. Boston Society Nat. History, Vol. XIII, with figure.
[Also printed and distributed separately as "Conchological Memoranda, No. 4," with figure.]

12. Description of a new species of Monoceros from California, with remarks on the distribution of North American species.

American Jour. Conchology, Vol. VII, 1871, 5 pp. with figure.

13. Descriptions of new California Shells.

American Jour. Conchology, Vol. VII, 1871, 2 pp. with figures.
[Preliminary descriptions of the species described in the two preceding papers were printed May 18, 1871, and distributed as "Conchological Memoranda, No. VI."]

14. Description of a New Species of Veronicella from Nicaragua.

Proceed. Boston Society of Natural History, 1871.
[Also separately as "Conchological Memoranda, No. VIII." October 11, 1871.]

15. On the Habitat and Distribution of the West American species of Cypraeidae, Triviidae, and Amphiperasidae, being corrections to Mr. Roberts' Catalogues of the "Porcellanidae and Amphiperasidae."

Proceed. Cal. Acad. Nat. Sciences, September 1, 1871.
[Also separately as "Conchological Memoranda, No. IX."]

16. Descriptions of New Species of Marine Mollusks from the Coast of Florida.

Proceed. Boston Society of Natural History, Vol. XV, Jan. 17, 1872.
[Also separately as "Conchological Memoranda, No. XI."]

17. Remarks on Pre-Historic Remains in Florida.

Proceed. Cal. Acad. of Sciences, March 1, 1872; also separately.

18. Description of a New Species of Mangelia from California.

Proceed. Cal. Acad. of Sciences, June 5, 1872, with figure.

19. Remarks on Marine Faunal Provinces on West Coast of America.

Proceed. Cal. Acad. of Sciences, July 1, 1872.

20. Descriptions of New Species of Shells from California.

Proceed. Cal. Acad. of Sciences, August 5, 1872, with figures.

21. Notes on Purpura canaliculata, Duclos.

Proceed. Cal. Acad. of Sciences, August 19, 1872.

22. A partial comparison of the Conchology of portions of the Atlantic and Pacific Coasts of North America.

Proceed. Cal. Acad. Sciences, October 7, 1872.
[Numbers 18, 19, 20, 21 and 22 included in "Conchological Memoranda, No. X," 7 pages and plate.]

23. On the Economic Value of certain Australian Forest Trees, and their cultivation in California.

Proceed. Cal. Acad. Sciences, July 1, 1872.
[Also in "California Horticulturist," Vol. II, and elsewhere; and separately in pamphlet, 11 pp. 8vo.]

24. The California Trivia and some points in its Distribution.

American Naturalist, Vol. VI, December, 1872, with figures.

25. Descriptions of a New Genus and two new species of Nudibranchiate Mollusks from the Coast of California.

Proceed. Cal. Acad. Sciences, April 7, 1873, with illustrations.

26. Descriptions of New Marine Mollusks from the West Coast of North America.

Proceed. Cal. Acad. Sciences, April 7, 1873, with plate.
[Numbers 25 and 26, also separately as "Conchological Memoranda, No. XII. Preliminary descriptions of most of the species described in No. 26 were printed and distributed in "Conchological Memoranda, No. VII," August 28, 1871.]

27. The Pectens, or Scallop Shells.

Overland Monthly (California), April, 1873.
[As referred to in History and Poetry; also remarks on their Anatomy.]

28. On Xylophagous Marine Animals.

California Horticulturist, etc., May, 1873, with figures.

29. Remarks on the Nudibranchiate, or Naked-gilled Mollusks.
California Horticulturist, July, 1873.

30. Aboriginal Shell Money.
Proceed. Cal. Acad. Sciences, July 7, 1873, with plate.
[Also separately, 8 pp., 8vo. and plate.]

31. Shells collected at San Juanico, Lower California, by William M. Gabb.
Proceed. Cal. Acad. Sciences, July 21, 1873.

32. Shells collected at Loreto, Lower California, by W. M. Gabb, in February, 1867.
Proceed. Cal. Acad. Sciences, July 21, 1873.
[Numbers 31 and 32, also separately.]

33. Aboriginal Shell Money.
Overland Monthly (San Francisco), Sept., 1873, with figures.
[Also separately, 10 pp., 8vo.]

34. Remarks on a New Alcyonoid Polyp, from Burrard's Inlet.
Proceed. Cal. Acad. Sciences, Feb. 3, 1873.
[Also separately, 8 pp.]

35. Descriptions of New Marine Shells, from the West Coast of Florida.
Proceed. Acad. Nat. Sciences of Philadelphia, July 22, 1873, with figures.

36. Description of a new species of Alcyonoid Polyp.
Mining and Scientific Press (San Francisco), August 9, 1873; also separately.

37. Description of a New Genus and species of Alcyonoid Polyp.
Proceed. Cal. Acad. Sciences, August 18, 1873; also separately.

38. Description of a New Genus and species of Alcyonoid Polyp.
Proceed. Cal. Acad. Sciences, August 18, 1873, with additions and plate. [Also separately, with plate.]

39. Prolific Joint Corn.
California Horticulturist, November, 1873.

40. Remarks suggested by Dr. J. E. Gray's Paper on the "Stick Fish," in "Nature," November 6, 1873.
Proceed. Cal. Acad. of Sciences, March 16, 1874. [Also separately.]

41. Remarks on the Upper Tuolumne Canyon.
Proceed. Cal. Acad. Sciences, September 15, 1873. [Also separately.]

42. On the Vitality of Certain Land Mollusks.
Proceed. Cal. Acad. Sciences, October 18, 1875, with plate. American Naturalist, Vol. XI, February, 1877.
[Also separately, with plate as "Conchological Memoranda, No. XIII."]

43. Description of New Fossil Shells from the Tertiary of California.
Proceed. Acad. Nat. Sciences of Philadelphia, 1875, with plate.

44. Aboriginal Shell Money.
American Naturalist, Vol. XI, June, 1877, with figures and plate.

45. Aboriginal Shell Ornaments, and Mr. E. A. Barber's Paper thereon.
American Naturalist, Vol. XI, August, 1877. [Also separately.]

REMARKS OF ROB'T E. C. STEARNS,

AND RESOLUTIONS OF THE

California Academy of Natural Sciences

UPON THE DEATH OF

ROBERT KENNICOTT.

California Academy of Natural Sciences.

REGULAR MEETING, OCTOBER 15th, 1866.

Mr. Stearns read the following:

It is my painful duty to inform the Academy of the decease of Robert Kennicott. The meager information received furnishes no particulars, further than that he died suddenly, in the month of May last, at Nulato Bay, in Russian America.

The services rendered to science by Mr. Kennicott are worthy of something more than a passing notice. In the month of May, in the year 1859, we find him starting upon a prolonged exploration of Russian America, under the auspices of the Smithsonian Institute, assisted by the University of Michigan, the Audubon Club of Chicago, and the Academy of Sciences of the same city. This exploration, including also a portion of the territory held by the Hudson's Bay Company, extended from May, 1859, to the date of his return in October, 1862. From the Annual Report of the Smithsonian Institute we learn that "the route traversed by Mr. Kennicott was from Lake Superior along the Kamenistiquoy River and Rainy and Winnipeg Lakes, up the Saskatchewan River to Cumberland House; thence nearly north to Fort Churchill, on English River, up the latter to Methy portage, at which point he first reached the head waters of the streams flowing into the Arctic ocean; thence along the Clear Water River and Athabasca Lake, down Peace River into Great Slave Lake, and along the Mackenzie River to Fort Simpson. At this place Mr. Kennicott spent a part of the first winter, making excursions up the Liard River to Fort Liard in autumn, and again on snowshoes in January. Before the close of the same winter he went up the Mackenzie to Big Island, and thence northwest to Fort Rae, near the site of old Fort Providence. From this point he traveled on the ice across Great Slave Lake to Fort Resolution, at the mouth of Peace River, where he spent the summer of 1860. He next descended the Mackenzie to Peel's River, and thence proceeded westward across the Rocky Mountains and down the Porcupine River to the Youkon, in the vicinity of which he spent the winter of 1860-61 and the summer of the latter year. The winter of

1861–2 was spent at Peel's River and LaPierre's house in the Rocky Mountains, and in traveling from this point to Fort Simpson and back to Fort Good Hope, on the Mackenzie. He left the last mentioned place on the first of June, 1862, and reached home in October. This enterprise terminated favorably, the explorer having returned richly laden with specimens, after making a series of observations on the physical geography, ethnology, and the habits of animals of the regions visited, furnishing materials of great interest to science."

Aside from the extensive collections in every department of natural history, the geographical information acquired by Mr. Kennicott was of the greatest importance.

In 1865, the Western Union Telegraph Company having determined to extend their wires so as to connect the old world and the new by an overland line passing through Russian America across Behring's Sea to Russia in Asia, and thence to the central cities of Europe, Mr. Kennicott's knowledge of the territory through which the proposed line was to pass made his services indispensable to the Company. He was sought out, and his coöperation at once secured. He entered upon this new labor, hoping not only to do whatever lay in his power to make this enterprise a success, but hoping also still further to serve the great cause that was so dear to him ; and while thus engaged in the enthusiastic performance of this self-imposed duty, in the prime of life, he has passed away.

The following resolutions were unanimously adopted.

Resolved, That the California Academy of Natural Sciences have learned with the deepest regret, of the untimely death of ROBERT KENNICOTT, and deplore the loss of one whose labors in the service of science entitle him to the grateful remembrance of his fellow men.

Resolved, That we extend our heartfelt sympathy to the family and friends of the deceased.

Resolved, That a copy of these Resolutions be forwarded to the family of the deceased, and to the Chicago Academy of Sciences, of which he was a prominent officer and energetic member.

CONCHOLOGICAL MEMORANDA,
NO. 2.

[From Proceedings Cal. Acad. Nat. Sci. August 20th, 1866.]

Mr. Stearns read the following:

Since my communication to the Academy of date of July 16th last, on the shells of Baulines Bay, additional specimens (4) of *Haliotis rufescens* have been found by Mr. Harford and Dr. Kellogg; also many specimens of *Katherina tunicata*, and one of *Mopalia Hindsii*; from between the umbos of very large specimens of *Mytilus Californianus*, collected by the same gentlemen, several specimens of *Barleia subtenuis* Carp.?

In addition to the above marine forms, the following species were found by the same parties in a gulch at Belvidere Ranch, not far from Capt. Morgan's house, south side of Baulines Bay: *Helix Nickliniana*, *H. arrosa*, *H. infumata*, *H. Columbiana* (hirsute var.) and *H. Vancouverensis*. Also, near a small stream on the same ranch, *Bythinella Binneyi*, Tryon. The last named species had previously been found in this neighborhood by Rev. J. Rowell.

[From Proceedings Cal. Acad. Nat. Sci. November 5th, 1866.]

Mr. Stearns exhibited several specimens of *Acmæa asmi*, Midd., collected by himself at Baulines Bay; remarking that upon a recent trip to the locality named, he found this species exceedingly numerous, attached to *Chlorostoma funebrale*, A. Ad.; that he had not seen in a single instance this Acmæa upon the rocks. Mr. Stearns further submitted for the inspection of the Academy, a specimen of *Haliotis Cracherodii*, which he had collected alive last month, on the rocks near the outlet of Lobitas Creek into the ocean. The Haliotis had been attacked by a pholad, probably *Navea Newcombii*, and had defended itself by adding coating upon coating of nacre, as the Navea progressed, until a large knob or protuberance had been created in the interior of the shell. From a partial examination of the borer, a specimen of which he had dug out from another portion of the same Haliotis, he believed it to belong to the species recently described by Mr. Tryon, viz: *Navea Newcombii*.

[From Proceedings Cal. Acad. Nat. Sci. December 3d, 1866.]

Mr. Stearns exhibited specimens of *Petricola carditoides* and *Pholadidea ovoidea*, in unusually hard serpentine, collected by himself at Fort Point, San Francisco.

[From Proceedings Cal. Acad. Nat. Sci. March 4th, 1867.]

Mr. Stearns read the following communication, prefacing it with some remarks on the hibernation and æstivation of land shells:

Remarkable Instance of Vitality in a Snail.

In that invaluable work to the conchological student, entitled "Recent and Fossil Shells," by S. P. Woodward, pp. 18 and 19, reference is made to certain genera and species of land shells, and several instances are cited proving the remarkable vitality of these comparatively insignificant animals, and their ability to exist for great lengths of time without food.

Particular mention is made of a specimen of the snail *Helix desertorum*, which was affixed to a tablet in the British Museum, March 25th, 1846, and upon the 7th of March, 1850, it was observed that the animal must have come out of the shell, as the paper was discolored in the attempt to get away, but finding escape impossible, it had withdrawn inside of the shell and closed the aperture with the usual glistening film, which led to its immersion in tepid water and marvelous recovery. It will be noticed that this period embraced nearly four years.

A more remarkable case has come under my observation, which is worthy of mention.

Dr. Veatch, a member of this Academy, visited Cerros or Cedros Island, opposite the west coast of Lower California, and upon his return, in the year 1859, brought home, among other shells, a species of Helix, supposed to be new, described by Dr. Newcomb, of Oakland, and to which the latter gave the name of *Helix Veatchii*; many specimens of this species were obtained, and some of them were given by Dr. Veatch to the late Thomas Bridges. Mr. Bridges died in September, 1865, and in December of the same year a portion of his collection passed into my hands, including the same specimens of *Helix Veatchii* to which I have before alluded. Judge of my surprise, when one day, upon a careful examination, I detected a living specimen, which, after being placed in a box of moist earth, in a short time commenced crawling about, apparently as well as ever. Fearing from its activity that by some accident it might crawl away, and I might thus lose it, after a fortnight's furlough from its long imprisonment, I placed it in a pill-box, marking the date of its reimprisonment upon the cover, in order that at some future time I may examine it, and ascertain possibly, if it does not outlive *me*, how long a snail can live without rations.

Here is an instance of a snail living at least *six* years—in Californian parlance, without a single "square meal."

[From Proceedings Cal. Acad. Nat. Sci. March 18th, 1867.]

Mr. Stearns made the following remarks as to the true habitat of *Helix Ayresiana*, Newc.:

On page 103, Vol. II, of the Academy's Proceedings, may be found, under date of March 18th, 1861, the description by Dr. W. Newcomb of a *Helix H. Ayresiana*, the habitat of which was, as I learn from Dr. N., doubtfully assigned at that time to "Northern Oregon." Recently Dr. Newcomb has himself detected it on Santa Cruz Island, off the Coast of California, near Santa Barbara.

[From Proceedings Cal. Acad. Nat. Sci. April 1st, 1867.]

Mr. Stearns presented the following papers:

Shells collected at Santa Barbara by W. Newcomb, M.D., in January, 1867.

BY ROBERT E. C. STEARNS.

1. Pholadidea penita, Conr.
2. ——— ovoidea, Gould.
3. Saxicava pholadis, Linn.
4. Platyodon cancellatum, Conr.
5. Cryptomya Californica, Conr.
6. Schizothaerus Nuttalli, Conr.
7. Neaera pectinata, Carp.
8. Clidiophora punctata, Conr.
9. Thracia curta, Conr.
10. Lyonsia Californica, Conr.
11. Mytilimeria Nuttalli, Conr.
12. Solen sicarius, Gould.
13. Solecurtus Californianus, Conr.
14. Machaera patula, Dixon.
15. Sanguinolaria Nuttalli, Conr.
16. Macoma secta, Conr.
17. ——— yoldiformis, Carp.
18. ——— nasuta, Conr.
19. ——— inconspicua, Br. & Sby.
20. Mera modesta, Carp.
21. Tellina Bodegensis, Hds.
22. Cooperella scintillaeformis, Carp.
23. Semele decisa, Conr.
24. ——— rupium, Sby.
25. Cumingia Californica, Conr.
26. Donax Californicus, Conr.
27. Standella planulata, Conr.
28. Amiantis callosa, Conr.
29. Pachydesma crassatelloides, Conr.
30. Psephis tantilla, Gould.

31. Chione succincta, Val.
32. Tapes staminea, Conr.
33. Saxidomus aratus, Gould.
34. Rupellaria lamellifera, Conr.
35. Petricola carditoides, Conr.
36. Chama exogyra, Conr.
37. Cardium quadragenarium, Conr.
38. Lazaria subquadrata, Carp.
39. Lucina Californica, Conr.
40. Diplodonta orbella, Gould.
41. Kellia Laperousii, Desh.
42. Mytilus Californianus, Conr.
43. Modiola capax, Conr.
44. —— recta, Conr.
45. Adula falcata, Gould.
46. —— stylina, Carp.
47. Pecten latiauritus, Conr.
48. Janira dentata, Sby.
49. Hinnites giganteus, Gray.
50. Ostrea lurida var. rufoides, Gould.
51. —— —— var. expansa, Carp.
52. Bulla nebulosa, Gould.
53. Haminea virescens, Sby.
54. Tornatina cerealis, Gould.
55. Dentalium hexagonum, Sby.
56. Mopalia muscosa, Gould.
57. —— ——?
58. Acanthopleura scabra, Rve.
59. Ischnochiton Magdalensis, Hds.
60. Nacella insessa, Hds.
61. —— depicta, Hds.
62. —— paleacea, Gould.
63. —— vernalis, Dall (mss.)
64. Acmæa patina, Esch.
65. —— persona, Esch.
66. —— scabra, Nutt. Rve.
67. —— spectrum, Nutt, Rve.
68. Lottia gigantea, Gray.
69. Scurria mitra, Esch.
70. Rowellia radiata, Cooper.
71. Clypidella bimaculata, Dall (mss.)
72. Fissurella volcano, Rve.
73. Glyphis aspera, Esch.
74. Lucapina crenulata, Sby.
75. Haliotis Cracherodii, Leach.
76. Phasianella compta, Gould.

77. —— pulloides, Carp.
78. Pomaulax undosus, Wood.
79. Trochiscus Norrisii, Sby.
80. Chlorostoma funebrale, A. Ad.
81. —— aureotinctum, Fbs.
82. Calliostoma canaliculatum, Mart.
83. —— costatum, Mart.
84. Crepidula lingulata, Gould.
85. —— excavata, Brod.
86. —— navicelloides, Nutt.
87. —— —— var mammaria, Gould.
88. —— —— var. explanata, Gould.
89. Hipponyx cranioides, Carp.
90. —— tumens, Carp.
91. —— serratus, Carp.
92. Serpulorbis squamigerus, Carp.
93. Turritella Cooperi, Carp.
94. Cerithidea sacrata, Gould.
95. Bittium filosum, Gould.
96. Litorina planaxis, Nutt.
97. —— scutulata, Gould.
98. Lacuna variegata, Carp.
99. —— unifasciata, Carp.
100. —— solidula, Loven.
101. Rissoa acutelirata, Carp.
102. Luponia spadicea, Gray.
103. Trivia Californica, Gray.
104. —— Solandri, Gray.
105. Erato vitellina, Hds.
106. Surcula Carpenteriana, Gabb.
107. Drillia inermis, Hds.
108. —— torosa, Carp.
109. —— mœsta, Carp.
110. Conus Californicus, Hds.
111. Odostomia sp.
112. Chemnitzia torquata, Gould.
113. Scalaria Indianorum, Carp.
114. —— subcoronata, Carp.
115. Cerithiopsis assimilata, C. B. Ad.
116. Lunatia Lewisii, Gould.
117. Ranella Californica, Hds.
118. Mitra maura, Swains.
119. Volvarina varia, Sby.
120. Olivella biplicata, Sby.
121. Nassa fossata, Gould.
122. —— perpinguis, Hds.

123. —— mendica, Gould.
124. —— Cooperi, Fbs.
125. —— tegula, Rve.
126. Amycla carinata, Hds.
127. —— tuberosa, Carp.
128. Amphissa corrugata, Rve.
129. Purpura crispata, Esch.
130. —— triserialis, Blve.
131. —— saxicola, Val.
132. Monoceros engonatum, Conr.
133. Ocinebra interfossa, Carp.
134. Cerastoma Nuttalli, Conr.
135. Muricidea Barbarensis, Gabb.

Dr. Newcomb also visited Santa Cruz Island, and collected the following species:

1. Waldheimia Grayi, Davidson.
2. Saxicava pholadis, Linn.
3. Semele decisa, Conr.
4. Acmaea scabra, Nutt.
5. Lottia gigantea, Gray.
6. Rowellia radiata, Cooper.
7. Haliotis Cracherodii, Leach.
8. —— corrugata, Gray.
9. Pomaulax undosus, Wood.
10. Trochiscus Norrisii, Sby.
11. Chlorostoma gallina, Fbs.
12. —— funebrale, A. Ad.
13. Chlorostoma aureotinctum, Fbs.
14. Trivia Californica, Gray.
15. Conus Californicus, Hds.
16. Amycla tuberosa, Carp.
17. Monoceros engonatum, Conr.
18. Cerastoma Nuttalli, Conr.
19. Muricidea incisus, Brod.
20. Trophon triangulatus, Carp.
21. Fusus ambustus, Gould.
22. Argonauta Argo, Linn.
23. Helix Ayresiana, Newc.

List of Shells collected at Purissima and Lobitas, California, October, 1866.

BY ROBERT E. C. STEARNS, CURATOR OF CONCHOLOGY, CAL. ACAD. NAT. SCIENCES.

"Purissima" and "Lobitas" are the names of two creeks situated a few miles south of Spanish Town, in San Mateo County. Near the points where these streams empty into the ocean are small beaches and groups of flat rocks left bare at low tide, limited, however, in extent, as the shore in the neighborhood is for the most part exceedingly bold and precipitous, the ocean at ordinary high water beating against the base of the cliffs.

Dr. Newcomb and myself visited the localities at the period above mentioned, and collected the following species from among the drift or upon the rocks:

1. Waldheimia Grayi, Davidson.
2. Navea Newcombii, Tryon.
3. Zirphaea crispata, Linn.
4. Pholadidea penita, Conr.
5. —— ovoidea, Gould.
6. Netastomella Darwinii, Sby.
7. Parapholas Californica, Conr.
8. Saxicava pholadis, Linn.
9. Platyodon cancellatum, Conr.
10. Schizothaerus Nuttalli, Conr.
11. Lyonsia Californica, Conr.
12. Mytilimeria Nuttalli, Conr.
13. Macoma inconspicua, Br. & Sby.
14. Standella falcata, Gld.
15. Tapes staminea, Conr.
16. —— ruderata, Desh.

17. Tapes diversa, Sby.
18. Rupellaria lamellifera, Conr.
19. Petricola cardiloides, Conr.
20. Lazaria subquadrata, Cpr.
21. Kellia Laperousii, Desh.
22. Mytilus Californianus, Conr.
23. —— edulis, Linn.
24. Modiola fornicata, Gld.
25. Adula falcata, Gld.
26. —— stylina, Cpr.
27. Hinnites giganteus, Gray.
28. Placunanomia macroschisma, Desh
29. Doris albopunctata, Cooper.
30. Cryptochiton Stelleri, Midd
31. Katherina tunicata, Wood.
32. Tonicia lineata, Wood.
33. Mopalia muscosa, Gld.
34. —— Hindsii, Gray.
35. —— lignosa, Gld.
36. Acanthopleura scabra, Rve.
37. Nacella vernalis, Dall (mss.)
38. —— instabilis, Gld.
39. Acmæa patina, Esch.
40. —— pelta, Esch.
41. —— Asmi, Midd.
42. —— persona, Esch.
43. —— spectrum, Nutt.
44. Lottia gigantea, Gray.
45. Scurria mitra, Esch.
46. Glyphis aspera, Esch.
47. Clypidella callomarginata, Cpr.
48. —— binmaculata, Dall (mss.)
49. Haliotis Cracherodii, Leach.
50. —— rufescens, Swains.
51. Chlorostoma funebrale, A. Ad.
52. —— brunneum, Phil.
53. —— Pfeifferi, Phil.

54. Calliostoma canaliculatum, Mart.
55. —— costatum, Mart.
56. Margarita pupilla, Gould.
57. Crepidula adunca, Sby.
58. —— navicelloides, Nutt.
59. —— var. nummaria, Gld.
60. —— var. explanata, Gld.
61. Hipponyx cranioides, Cpr.
62. Litorina planaxis, Nutt.
63. —— scutulata, Gld.
64. Lacuna porrecta, Cpr.
65. —— unifasciata, Cpr.
66. Isapis obtusa, Cpr.
67. Erato vitellina, Hinds.
68. Drillia torosa, Cpr.
69. Scalaria Indianorum, Cpr.
70. —— subcoronata, Cpr.
71. Opalia borealis, Gld.
72. Velutina prolongata, Cpr.
73. Olivella biplicata, Sby.
74. —— intorta, Cpr.
75. Nassa fossata, Gld.
76. —— perpinguis, Hds.?
77. —— mendica, Gld.
78. Amycla gausapata, Gld.
79. Amphissa corrugata, Rve.
80. Purpura crispata, Chem.
81. —— var. septentrionalis, Rve.
82. Purpura saxicola, var. ostrina, Gld.
83. Monoceros engonatum, Conr.
84. Ocinebra lurida, Midd.
85. —— var aspera, Baird.
86. —— var. munda, Cpr.
87. —— interfossa, Cpr.
88. Cerostoma foliatum.
89. Octopus —— (n. s.?)

Narca Newcombii, alive in *Haliotis Cracherodii*, Nos. 4, 5, 6, 25, and 26 alive in soft shale between tide marks. *Doris albopunctata*, two specimens alive on rocks near low water mark. Of the Chitons, Nos. 31 and 36, particularly abundant; of the others named several specimens obtained, also one or two species undetermined. 41, common, alive, on *Chlorostoma funebrale*. 45 and 46, several living specimens between tide marks. 47 and 48, I think, are distinct species; suggest *Lucapina*, but foramen nearly twice as large as in shells of the latter of same size, differing also in sculpture and weight of shell. 49, animal lives for a long time, and affixes itself tenaciously to the rocks after the shell is removed. 63 and 65, together living on rocks near high-water mark, and on eel grass in pools left by the tide. 89, perhaps young of Mr. Gabb's species *O. punctatus;* two living specimens, as yet undetermined, probably a new species.

CONCHOLOGICAL MEMORANDA.
NO. III.

(From Proceedings Cal. Acad. Nat. Sciences, December 2d, 1867.)

List of Shells collected at Bodega Bay, California, June, 1867.

BY ROBT. E. C. STEARNS, CURATOR OF CONCHOLOGY, CAL. ACAD. NAT. SCIENCES.

In pursuance of the idea mentioned in my paper on the shells of Baulines Bay, of examining the bays and coast to the north of San Francisco, I made a brief trip to Bodega Bay in company with my friend Dr. Newcomb, on the thirteenth of June, 1867. Most of the species enumerated were collected within a very limited area, between tide marks, at the extreme point of Bodega Head, as the arm of land is called, which extending in a southerly direction from the general line of the coast, incloses what is known as Bodega Bay. The bay itself is, for the greater part, left bare at low tide, and the flats then exposed, composed of sandy mud, contain abundance of the common bivalves of the coast, principally *Macoma*, (two species) and *Tapes*, in all its varieties; *Saxidomus gracilis* may also be found here in considerable quantities, and is at certain seasons dug by the Indians, together with the other so called "clams." At the spot where the principal portion of this collection was made, the outcropping rock is a coarse granite, upon which *Litorina planaxis* is found in great numbers. The limited time at my disposal, at the season when the trip was made, was only sufficient to admit of a brief, and therefore unsatisfactory reconnoissance; nevertheless, at least seventeen species were detected which have not heretofore been found (or reported) so far to the north. Many of these species I failed to find at Baulines, and some of them have not been reported north of the Bay of Monterey. At Baulines, the rocks are principally shales, and contain many species of pholads, which as will be seen by a glance at this list, if not entirely absent, must be rare at Bodega; the various "nestlers" which are found associated with the borers are also wanting; *Haliotis rufescens* is abundant upon the rocky islets off the coast, but not even a fragment of *H. Cracherodii* was met with.

1. Cryptomya Californica, Conr.
2. Schizothaerus Nuttalli, Conr.
3. Entodesma saxicola, Baird.
4. Mytilimeria Nuttalli, Conr.
5. Machera patula, Dixon.
6. Macoma secta, Conr.
7. —— nasuta, Conr.
8. Tellina Bodegensis, Hds.
9. Tapes staminea, Conr.‡
10. —— —— var. Petitii, Desh.‡
11. —— —— var. ruderata, Desh.‡
12. —— —— var. diversa, Sby.‡
13. Saxidomus gracilis, Gould.
14. Chama exogyra, Conr.
15. Cardium corbis, Mart.
16. Lazaria sub-quadrata, Cpr.
17. Kellia Laperousii, Desh.
18. Lasea rubra, Mont.

19. Mytilus Californianus, Conr.
20. ——— edulis, Linn.
21. Modiola fornicata, Cpr.*
22. ——— recta, Conr.*
23. Axinæa subobsoleta, Cpr.
24. Pecten hastatus, Sby.
25. Hinnites giganteus, Gray.
26. Placunanomia macrochisma, Desh.
27. Helix Nickliniana, Lea.
28. ——— Columbiana, Lea.
29. Cryptochiton Stelleri, Midd.
30. Tonicia lineata, Wood.
31. Mopalia Wossnessenskii, Midd.
32. ——— Merckii, Midd.
33. Kennerlyi var. Swanii, Cpr.
34. Trachydermon fallax, Cpr. (Mss.)
35. Nacella instabilis, Gould.
36. Acmæa patina, Esch.
37. ——— pelta, Esch.
38. ——— persona, Esch.
39. ——— scabra, Nutt.*
40. ——— spectrum, Nutt.
41. Scurria mitra, Esch.
42. Rowellia radiata, Cp.*
43. Glyphis aspera, Esch.
44. Clypidella callomarginata, Cpr.*
45. ——— bimaculata, Dall, (Mss.)*
46. Haliotis rufescens, Swains.*
47. Leptothyra sanguinea, Cpr.
48. Chlorostoma funebrale, A. Ad.
49. ——— brunneum, Phil.
50. Calliostoma costatum, Mart.
51. ——— annulatum, Mart.
52. Phorcus puliigo, Mart.
53. Margarita pupilla, Gould.

54. ——— acuticostata, Cpr.*
55. Crepidula adunca, Sby.
56. Hipponyx cranioides, Cpr.
57. ——— antiquatus, Linn.*
58. Bivonia compacta, Cpr.
59. Bittium filosum, Gould.
60. Littorina planaxis, Nutt.
61. ——— scutulata, Gould.
62. Lacuna porrecta, Cpr.
63. Trivia Californiana, Gray.*
64. Erato vitellina, Hds.*
65. Drillia incisa, Cpr.
66. Mangelia levidensis, Cpr.‡
67. Odostomia gravida, Gould.*
68. Scalaria subcoronata, Cpr.*
69. Opalia borealis, Gould.
70. Velutina lævigata, Linn.
71. Lunatia Lewisii, Gould.
72. Olivella biplicata, Sby.
73. ——— bœtica, Cpr.
74. ——— intorta, Cpr.*
75. Nassa fossata, Gould.
76. ——— mendica, Gould.
77. Amycla carinata var. Hindsii, Rve.
78. ——— gausapata, Gould.
79. Amphissa corrugata, Rve.
80. Purpura crispata, Chem.
81. ——— canaliculata, Duel.
82. ——— saxicola var. ostrina, Gld.
83. Ocinebra lurida, Midd.
84. ——— ——— var. aspera, Baird.
85. ——— interfossa, Cpr.
86. ——— ——— var. atropurpurea, Cpr.
87. Cerostoma foliatum, Gmel.

* The species marked with an asterisk, seventeen in number, have never before been reported from a locality so far north.

† Mangelia levidensis (teste J. G. Cooper) has not previously been detected at a point so far south; it has heretofore been credited to "Straits of Fuca, W. T." vide Geo. Survey Cat. 1867, by J. G. C.

‡ Tapes staminea and vars. were obtained at low water by digging from twelve to twenty inches deep, and together with Macoma secta and M. nasuta, were found in the same holes.

The Chitons above enumerated, have been compared with specimens recently (March, 1868) received labeled, from Dr. Carpenter of Montreal.

No. 39, Acmæa scabra; elevated dark colored specimens of this species with the characteristic sculpture sharply and well defined, were obtained in considerable numbers. Subsequently at Monterey I found occasional specimens displaying nearly the same elevation and of the same color as those from Bodega.

(From Proceedings Cal. Acad. Nat. Sciences, December 24, 1867.)

Shells collected by the U. S. Coast Survey Expedition to Alaska, in the year 1867.

BY ROBT. E. C. STEARNS, CURATOR OF CONCHOLOGY, CAL. ACAD. NAT. SCIENCES.

George Davidson, Esq., connected with the Coast Survey service of the United States, who commanded the scientific department of the Alaska Expedition, very kindly tendered positions on his staff to the following members of the Academy: Dr. A. Kellogg, as Surgeon and Botanist; Theodore A. Blake, as Geologist; and W. G. W. Harford, as General Collector, by whom the species here enumerated were collected. My acknowledgments are due to Dr. J. G. Cooper, of San Francisco, for assistance in determining species; also to Dr. William Stimpson of Chicago, for similar service in reference to the Buccinidæ.

1. Saxicava pholadis, Linn, var. arctica; Sitka, Bella Bella, Kodiak, Oonalaska.
2. Mya arenaria, Linn.; Kodiak.
3. Schizothaerus Nuttalli, Conr.; Sitka, Kodiak.
4. Machaera patula, Dixon; Kodiak, Oonalaska.
5. Macoma nasuta, Conr.; Kodiak.
6. Macoma inquinata, Desh.; Fort Simpson, Bella Bella, Kodiak, Spruce Isl.
7. Macoma inconspicua, Br. & Sby.; Fort Simpson, Chilchat, Kodiak, Spruce Isl.
8. Mera salmonea, Cpr.; Kodiak.
9. Standella planulata, Conr.; Kodiak, Oonalaska.
10. Tapes staminea, var. Petitii, Desh.; Fort Simpson, Chatham Sound.
11. Tapes staminea, var ruderata, Desh.; Fort Simpson, Carter's Bay, Sitka, Bella Bella, Kodiak, Spruce Isl. Oonalaska.
12. Saxidomus Nuttalli, Conr.; Ft. Simpson, Sitka, Carter's Bay, Kodiak.
13. Cardium corbis, Mart.; Sitka, Bella Bella, Carter's Bay, Kodiak.
14. Cardium blandum, Gould; Sitka, Kodiak, Oonalaska.
15. Serripes Groenlandicus, Chem.; Kodiak, Oonalaska.
16. Kellia Laperousii, Desh.; Oonalaska.
17. Lasea rubra, Mont.; Sitka.
18. Mytilus edulis, Linn.; Ft. Simpson, Carter's Bay, Bella Bella, Sitka, Kodiak, Spruce Isl.
19. Modiola modiolus, Linn.; Sitka, Oonalaska.
20. Modiolaria laevigata, Gray; Oonalaska.
21. Aximea septentrionalis, Midd.; Bella Bella.
22. Yoldia, n. s.?; Stomach of Halibut, Kodiak.
23. Acila castrensis, Hinds; Sitka.
24. Placunanomia macroschisma, Desh.; Kodiak, Oonalaska.
25. Helix Columbiana, Lea; Sitka, Chilchat River, 59° 9′ N
26. Helix Vancouverensis, Lea; Sitka, V Island.
27. Helix ruderata, Stud.; Oonalaska.
28. Helix fulva, Drap.; Sitka, Oonalaska.
29. Vitrina pellucida, Mull.?; Oonalaska.
30. Zua lubrica, Mull.; Sitka, Kodiak.
31. Siphonaria thersites, Cpr.; Fort Simpson.
32. Katherina tunicata, Wood; Sitka.
33. Tonicia lineata, Wood; Fort Simpson.

34. Tonicia submarmorea, Midd.; Fort Simpson.
35. Mopalia muscosa, Gould; Vancouver Island.
36. Mopalia Wossnessenskii, Midd.; Fort Simpson.
37. Mopalia Merckii, Midd.; Fort Simpson.
38. Acmaea patina, Esch.; Fort Simpson, Kodiak, Oonalaska, Sitka
39. Acmaea pelta, Esch.; Sitka, Kodiak, Oonalaska.
40. Scurria mitra, Esch.; Sitka.
41. Glyphis aspera, Esch.; Sitka.
42. Haliotis Kamschatkana, Jonas; Sitka.
43. Calliostoma costatum, Mart.; Sitka.
44. Margarita pupilla, Gould; Ft. Simpson, Bella Bella, Sitka, Oonalaska.
45. Margarita helicina, Mont.; Oonalaska.
46. Phorcus pulligo, Mart.; Sitka.
47. Crepidula navicelloides, Nutt.; Bella Bella.
48. Crepidula grandis, Midd.; Captain's Harbor, Kodiak, Oonalaska.
49. Bittium filosum, Gould; Ft. Simpson, Carter's Bay, Bella Bella, Sitka.
50. Littorina scutulata, Gould; Sitka.
51. Littorina Sitkana, Phil.; Chatham S'nd, Carter's Bay, Bella Bella, Sitka, Kodiak.
52. Lacuna solidula, Loven.; Oonalaska.
53. Isapis fenestrata, Cpr.; Oonalaska.
54. Trichotropis cancellata, Hds.; Fort Simpson, Sitka.
55. Natica clausa, Brod. & Sby.; Ft. Simpson, Kodiak, Oonalaska.
56. Lunatia pallida, Brod. & Sby.; Captain's Harbor, Oonalaska.
57. Priene Oregonensis, Redf.; Oonalaska.
58. Olivella boetica, Cpr.; Sitka.
59. Nassa mendica, Gould; Sitka.
60. Amycla gausapata, Gould; Ft. Simpson.
61. Amphissa corrugata, Rve.; Ft. Simpson, Carter's Bay.
62. Purpura crispata, Chem.; Fort Simpson, Bella Bella, Lawson's Harbor, Carter's Bay, Sitka.
63. Purpura canaliculata, Duel.; Chatham Sound, Carter's Bay, Bella Bella, Sitka, Kodiak, Spruce Isl., Oonalaska.
64. Purpura saxicola, Val.; Oonalaska.
65. Purpura saxicola var. fuscata, Fbs.; Fort Simpson, Carter's Bay, Bella Bella, Sitka.
66. Ocinebra interfossa, Cpr.; Carter's Bay, Bella Bella, Sitka.
67. Cerastoma foliatum, Gmel.; Bella Bella, Sitka.
68. Trophon multicostatus, Esch.; Oonalaska.
69. Trophon orpheus, Gould; Sitka.
70. Chrysodomus dirus, Rve.; Chatham Sound, Ft. Simpson, Carter's Bay, Bella Bella, Sitka.
71. Chrysodomus liratus, Mart.; Chilchat, Kodiak, Oonalaska.
72. Buccinum glaciale, Linn.; Oonalaska.
73. Buccinum polare, Gray; Captain's Harbor, Oonalaska.
74. *Buccinum cyaneum, Brug.; Kodiak.
75. Volutharpa ampullacea, Midd.; Fort Simpson, Bella Bella, Sitka, Kodiak, Oonalaska.

---

* In a note from Dr. Stimpson, he remarks in reference to this species: it "has not, as far as I am aware, as yet been reported from the Pacific."

CONCHOLOGICAL MEMORANDA. NO. 4.

[From the Proceedings of the Boston Society of Natural History, Vol. XIII.]

ON A NEW SPECIES OF PEDIPES FROM TAMPA BAY, FLORIDA. BY ROBERT E. C. STEARNS.

Pedipes naticoides Stearns. Shell resembling a tiny Natica; imperforate, globose, translucent, pale horn color; spire short, apex obtuse; whorls four to four and a half, slightly flattened above; the upper whorls moderately elevated; body whorl nine tenths the length of the shell; suture strongly defined; surface ornamented with fine depressed revolving lines, crossed obliquely and regularly by sharply developed lines of growth; aperture longitudinal, suboval; the middle portion of the outer lip moderately tuberculately thickened within, and slightly pressed inwards, giving a somewhat angular outline to

the upper part of its edge; parietal wall covered with shining callus and furnished internally with a strongly developed ridge or plait, which culminates in a prominent sub-acute tooth, projecting in the line of its obliquity two fifths of the width of the aperture; columella showing two rather obtuse teeth or folds, the upper being the largest, with a sinuous sulcation at their bases, parallel to the outline of the columella, causing, together with the folds, an appearance resembling the thread of a screw, or the plaits in Cancellaria.

Long. .11 inch. Lat. .08 inch.

Habitat: Littoral zone, Rocky point, Tampa Bay, western shore of Florida; found with other small species of mollusks upon the under side of clumps of "Coon oysters" at low water line. Two specimens, living, one adult, the other not quite developed.

This well marked species is the first of the genus found upon the eastern side of the Continent, and the fourth thus far detected in North America.

The late Prof. C. B. Adams obtained a species at Panama, *P. angulatus*, Mr. W. G. Binney described another, *P. liratus*, from Cape St. Lucas, and Dr. J. G. Cooper a third, *P. unisulcata*, from San Pedro, California; the species above described is more globose, and more delicate than either of the others.

For the excellent figure of *P. naticoides* I am indebted to the kindness of my friend, Mr. E. S. Morse.

Conchological Memoranda, No. VI.

May 18, 1871.

PRELIMINARY DESCRIPTIONS OF NEW SPECIES FROM THE WEST COAST OF AMERICA.

BY R. E. C. STEARNS.

Monoceros paucilirata; Stearns.

Shell moderately elevated, whorls 4-6; body whorl four-fifths the total length, angulated above and excavated between the angle and the suture; a sharp groove behind the tooth. Upper whorls cancellated, nucleus smooth. Aperture elongate, purple brown in the throat; outer lip sharp, yellowish, internally denticulated, with a prominent tooth at its outer edge. Columella purple, canal short, umbilicus nearly covered by the columellar callus. Siphonal fasciole strong. Externally painted with longitudinal broad black and narrow whitish streaks, interrupted by the white dental groove and three or four narrow yellowish revolving carinæ, which, except the keel, are inconspicuously elevated. Lon, .55; Lat, .33 in. Habitat—Coronado Islands, off San Diego, California. Hemphill, three specimens.

Ocinebra circumtexta; Stearns.

Shell ovate, solid, sub-turreted, of five convex whorls. Upper whorls cancellated; body whorl traversed by about 11 roughly-rounded revolving costæ, more or less tuberculated at the intersection of the longitudinal ribs, and marked with fine incremental striæ. Last whorl three-fourths the length of the shell; outer lip thickened internally denticulate, external edge crenulated. Columella excavated, light purple or purplish brown; canal short, open or closed in specimens of equal size. Umbilicus obsolete; surface of whorls with faint irregular longitudinal costæ. Color dingy white, with two interrupted black or dark brown bands. Lon, .85; Lat, .5 in. Habitat—Monterey, California; Hemphill, Harford, Gordon, and Stearns, sixteen specimens, mostly immature.

Ocinebra gracillima; Stearns.

Shell small, solid, fusiform, slender; spire subacute; whorls 6-7; body whorl about two-thirds the whole length. Upper part of whorls subangulate, aperture about as long as the spire. Outer lip thickened internally; white, with four prominent denticles. Columellar lip excavated, callous, with a purplish stain showing through the enamel. Canal moderate, closed. Surface smooth, with numerous fine whitish revolving costæ, dotted with brown, the interspaces near the outer lip with brown linear markings. Upper whorls longitudinally nodosely ribbed. General color olivaceous, with patches of yellow. Lon, .5; Lat, .25 in. Habitat—San Diego, California, 10 fms.; Hemphill.

The above are merely preliminary to detailed descriptions hereafter to be published in the American Journal of Conchology.

DESCRIPTION OF A NEW SPECIES OF MONOCEROS. FROM CALIFORNIA, WITH REMARKS ON THE DISTRIBUTION OF THE NORTH AMERICAN SPECIES.

BY ROBERT E. C. STEARNS.

M. PAUCILIRATA, Stearns. Pl. 14, fig.16.

M. paucilirata, Stearns, Prel. Descr. May 18, 1871.

Description. Shell small, ovate, spire moderately elevated, subacute; whorls four to six; body whorl four-fifths the length of the shell; upper portion of same angulated, and excavated between the angle and the suture, and anteriorly broadly but not deeply grooved; upper whorls rudely cancellated except the apex, which is nearly smooth; aperture ovate, purplish-brown; columella flattened, enameled, purple, brown, or blackish, sometimes showing all of these colors; outer lip simple, acute, internally denticulated, whitish or yellowish near the edge, with a single prominent tooth at its anterior margin; canal short, slightly recurved; umbilicus nearly concealed; body whorl traversed spirally by four to five narrow ribs placed nearly equidistant, and longitudinally marked by irregular varix and fine incremental lines; shell yellow or yellowish-white, with large angular spots of black; the transverse costæ and the more prominent of the longitudinal lines are of the lighter color, and the largest specimen exhibits faint transverse sculpture between the prominent ribs.

Length of largest specimen, ·55. Width ·33 inch.

Habitat, Coronado Islands, off San Diego, California.

Three specimens of this well-marked species, quite distinct from others of the genus, were collected by Mr. Hemphill.

A comparison of the above shell with *immature* specimens of *Monoceros lugubre*, Sby., shows the former to be a connecting link between the upper Californian and more southern forms. *M. lugubre* reaches northerly nearly to San Diego, being found at Los Todos Santos Bay, where numbers of specimens were collected by Mr. Hemphill. Fossil specimens of the latter have been collected from the post-pliocene formation at Santa Barbara Island, Cal. (*vide* Geol. Survey of Cal., Paleon. Vol. II, p. 75).

The geographical range of the North American species of this genus, which has no representative upon the Atlantic coast of the continent, is as follows: Commencing at the north, we find *M. engonatum*, Conr., with its northern limit at Baulines (or Bolinas) Bay, which is about twelve miles north of the entrance to the Bay of San Francisco, where I have collected great numbers of specimens, generally of small size, upon the shales, between ordinary tide-marks: it extends southerly to San Diego, a range of nearly four hundred miles, where it is represented by a local varietal form, which is the *M. spiratum* of Blainville. *M. engonatum* is by far the most common of the northern species; in one instance I collected not less than two thousand specimens, and I find it quite generally distributed along the coast. The finest specimens, averaging twice the size of the Baulines shells, may be found at Lobitas, Point Año Nuevo, thence to Monterey and some distance southward. In the post-pliocene at San Pedro, Cal., it is found fossil, (Geol. Survey of Cal., Paleon. Vol. II, 75.) It is the *M. unicarinatum*, Sby. (*vide* Conch. Illustr. genus Monoc., fig. 5,). Sowerby's name would apply very well to the var. *spiratum*, but his figure distinctly represents the general form.

M. lapilloides, another of Mr. Conrad's species, a handsome and more globose shell than *M. engonatum*, is much less abundant and far more restricted in range; it equals *M. "punctatum*, Gray + *brevidens*, Conr." (*vide* Cpr. Supp. Rep. 1863), and *M. punctulatum*, Gray. *vide* Sby.'s Conch. Illustr. genus Monoc., fig. 3, where it is well figured. It has not, to my knowledge, been collected north of Monterey, which place, if numbers of individuals are to be considered, seems to be its specific centre. Cooper, in his Geog. Cat. credits it to Catalina Island, which is perhaps its southern limit, but neither of the collections made by Hepburn, Newcomb, Hemphill or Harford, contained a single specimen either from Santa Barbara or San Diego. All of the collections above referred to were examined and noted by me, as well as a collection made at Catalina Island by Mr. Cummings, of San Francisco, which contained many rarities, but not a specimen of this species. I therefore infer that, though reported from Catalina Island, it is seldom met with even there. This conclusion is further confirmed by the absence of the species from several parcels of shells from said island received by me at various times from my friend W. M. Cubery, Esq.

Next in order, and to the south, comes the species described by me in this paper, which, as suggested, connects the northern with the semitropical forms.

The form to which I have given the name of *pauciliratum* seems to have escaped the keen search of many experienced collectors, as not a single specimen of it was detected by either of the gentlemen above named, and it was not discovered until Mr. Hemphill visited the small islands off San Diego, known as the "Coronados." As the mainland has been pretty well examined along here, it is probably an insular species, and a thorough investigation of the islands may prove it to be quite common. Its distribution may be somewhat eccentric, like that of the fine *Purpura planospira*, Lam., which is seldom found upon the mainland, and confines itself, with aristocratic exclusiveness, to Socorro Island, one of the islands off the west coast of Mexico, known as the Revillagigedos.

In the third edition of Dr. Jay's catalogue (1839) a species of Monoceros, *M. plumbeum*, Kiener, is credited to Upper California; to what shell this refers I am not aware. (*Pseudoliva* [*Buccinum*] *plumbea*, Desh. W. H. Dall). Kiener's species does not appear in the fourth edition of said work, but " *M. plumbeum*, Chemn.," the habitat of which is given as " Africa Merid."

Inhabiting a long reach of coast, extending the entire length upon the ocean side of the peninsula of Lower California, a distance of nearly eight hundred miles, as well as at Cedros and other islands we find the next species in order, *Monoceros lugubre* of Sowerby; it is also found up the Gulf of California, and is reported from Guaymas on the eastern shore. In many places it is exceedingly numerous, and is perhaps the commonest species in collections; specimens from different points vary considerably, the outer coast or ocean side shells being the most robust, and the Gulf specimens more elongated. I did not obtain a specimen at Acapulco nor hear of its being found so far to the south.

It is not reported in the Mazatlan catalogue of Reigen's collection; in Kiener it is credited to " Peru and California," but as Kiener, Reeve, Sowerby and many others of the European authors seem, to use a California phrase, to have " gone it blind" on habitat, in questions of geographical distribution, their statements of localities must be taken with great caution. In the Tankerville catalogue it is referred to as *M. cymatum*, Shy., in H. and A. Ad. Genera, as *M. cymatium*, Soland., and neither of Conrad's names are given for the California species by the last named authors. *M. engonatum*, Conr., is enumerated in the list under Blainville's name "*spirata*" and *M. lapilloides*,

Conr., appear as the "*M. punctata*, Gray." It is highly improbable that *lugubre* occurs on the coast of Peru. In the large collection made by the late Thomas Bridges, upon the South American coast, it was not found, and is not mentioned in the Panama Catalogue by Prof. C. B. Adams; Kiener's "Peru" is undoubtedly an error.

The next species in order is the *Monoceros muricatum*, Rve., which is the *Purpura muricata*, Gray; it is also enumerated under the latter name as a *Purpura*, in Smithsonian check list, by Dr. Carpenter, who refers to it in his Mazatlan Cat., p. 476, as follows. "This shell rests its claim to a place in the genus *Monoceros* on a projecting wave in the labrum, between the canal and the first costal depression." The finely developed specimens in my own collection, as well as many that I have seen elsewhere, justify me in supporting the "claim" of this shell "to a place in the genus *Monoceros*," as the claim is well founded; the "projecting wave," to quote Dr. Carpenter, is in my specimens developed into a horn of sufficient prominence to enable the shell to enter the genus *Monoceros* on its own hook. I will state that, in many large specimens that have come under my observation, the horn can hardly be considered as anything more than a "projecting wave." Individuals of the same species, and in all of the species, vary exceedingly in this respect.

M. muricatum, though not a rare shell, is not common. In the Reigen collection it is reported as rare at Mazatlan; I have received many specimens at various times from different parties, generally immature, however, and their exact habitat, or that of the finely developed specimens above mentioned, I am unable to give. It was collected by Major William Rich, in Lower California, and in the collection made by Lieut. T. P. Green, it is reported from San Juan, which is no doubt correct, though the small collection made there by Prof. Gabb did not contain it. This species appears in collections named as *Purpura* or *M. muricata, muricatum* and *tuberculatum*.

In Adams' genera it is catalogued as "*M. tuberculata*, Gray;" there is a good picture of it in Sby's. Conch. Illustr., fig. 9, and in Chenu's "Manuel," on p. 169, fig. 828, and *M. tuberculatum* fig. 831.

In the Mazatlan Cat., on the authority of Cuming, one habitat is given as "St. Elena," which is on the coast of Guayaquil, latitude about 2° south, but no mention is made of it in Prof. Adams' Panama shells, and I did not find it in the Bridges' collection. There is a long reach of coast extending from Acapulco to Panama which is, following the bends and curves of the shore line, not less than fourteen hundred miles in length, of which, with the exception of a few places, but little is known; somewhere upon this long line of beach or rock, in some of the numerous bays, it may quite likely be abundant.

The last and most southern of the North American species, a shell plentifully distributed within certain limits, is the *M. brevidentatum* of Wood, which is catalogued by the Adams' in their Genera, Vol. I, p. 131, as "*cornigera*, Blainv.," though Wood's name has, according to Prof. C. B. Adams' dates, four years priority; the more common of the synonyms by which this species is known are *M. ocellata*, Kiener, and *M. maculata*, Gray.

Numerous blunders have been made in the habitat of this species, and it has been reported from San Francisco! and Monterey! also from Mazatlan, all of which are erroneous, as it is undoubtedly confined to the zoological province of Panama. Carpenter makes no mention of it in the Reigen collection and it is not reported by him in the Xantus list of Cape St. Lucas shells.

In the large quantities of material from the Gulf of California that I have overhauled and examined, I have never met with a specimen of it, and I have yet to learn from any authentic source of its occurrence north of San Juan del Sur Nicaragua.

I have not attempted a systematic synonymy of the species herein named; the original works not being accessible I could only quote them indirectly, and therefore the result at best would have been unsatisfactory. I have, however, given a sufficient reference to the more common synonyms, for the reader to understand me, and to recognize the shells to which I have had occasion to refer.

DESCRIPTIONS OF NEW CALIFORNIA SHELLS.

BY ROBERT E. C. STEARNS.

OCINEBRA, Leach.

O. GRACILLIMA, Stearns. Pl. 14, fig. 14.

O. gracillima, Stearns, Prel. Descr. May 18, 1871.

Description.—Shell small, solid, ovately fusiform, rather slender; spire elevated, subacute; whorls six—seven; body whorl about two-thirds of the length of the shell; upper portion of whorls more or less angulated; aperture ovate, about the same length as the spire; outer lip thickened, internally white, with four prominent denticles; columellar lip somewhat excavated, incrusted, with a purplish stain showing through the enamel; canal closed, moderately recurved; surface covered with a slight yellowish epidermis and numerous revolving costæ of a whitish hue, finely dotted with reddish brown, and the intercostal sulcations near the edge of the outer lip with linear markings of the same color; (one specimen shows brownish blotches upon the middle of the body whorl.) This shell is also longitudinally nodosely ribbed, the ribs decreasing in prominence as the whorls enlarge.

Length, ·50; width, ·25 inch.

Habitat, San Diego and vicinity, California.

One specimen, an adult, dredged in (10 fathoms) San Diego bay; one adult and five juniors ten miles above San Diego; also one adult specimen at Dead Man's Island, near San Pedro.

For this exceedingly pretty species we are indebted to Henry Hemphill, Esq.

O. CIRCUMTEXTA, Stearns. Plate 14, fig. 15.

O. circumtexta, Stearns, Prel. Descr. May 18, 1871.

Description.—Shell ovate, solid, spire subacute, in some specimens subturrited; whorls five, convex or moderately flattened above, upper whorls cancellated; body whorl nearly three-fourths of the length of the shell; outer lip thickened, internally denticulated, its external margin crenulated; columella excavated; aperture ovate, light purple to purplish brown; canal short, moderately curved, in some specimens closed, in others, of equal or larger size, open; umbilicus nearly obsolete; surface traversed by about fourteen roughly rounded revolving costæ, more or less varicose or tuberculated at the intersection of the longitudinal ribs and varical ridges; also marked with fine incremental striæ, the latter being more conspicuous in the intercostal sulcations; color white, somewhat dingy, with two interrupted zones of black or dark brown on the body whorl.

Number of specimens examined sixteen, mostly immature; the dimensions of the largest are respectively:

Length,	·85	·80	·75 inch.
Breadth,	·50	·47	·45 inch.

Habitat, Monterey, California, where it has been collected by Messrs. Hemphill, Harford, Gordon and myself.

This shell is *Ocinebra* var. *d* of Carpenter (MSS.), and 762 (in part) of Cooper's Geog. Cat. Moll.

The above species varies considerably in robustness, length of spire and development of sculpture: it is exceedingly characteristic in its markings, and easily distinguished from all others belonging to the Californian group of the *Ocinebræ*; it is not numerous in individuals, and appears to be exceedingly limited in its distribution.

Since writing the above, additional specimens have been forwarded to me by Mr. Harford, collected by him at the Island of San Miguel, off Santa Barbara, Cal. These latter specimens show a much more prominent longitudinal sculpture than the Monterey specimens, are generally more rugose, and one specimen is ashen white and wants the interrupted color bands.

NOTES ON THE LINGUAL DENTITION AND JAWS OF TERRESTRIAL MOLLUSCA. NO. 3.

BY THOMAS BLAND AND W. G. BINNEY.

In the following pages we have not considered it necessary to describe fully the lingual dentition of those species which agree with the usual type of dentition in their respective genera. In the *Helicidæ* we have given more particular notes on the marginal teeth, because they are not so constant in their characters as the central and lateral teeth.

ZONITES CAPSELLA, Gould.

Lingual membrane broad, not very long. Teeth as in the species of *Hyalina* figured by us in Land and Fresh-water Shells, Part I. The laterals and centrals are about equal in size, the former very few in number, apparently three only. Marginals numerous, large, decreasing in size as they pass off laterally, and quite separated near the outer margin of the membrane.

Specimen from Lexington, W. Virginia, received from Professor McDonald.

ZONITES LIGERUS, Say.

Lingual membrane as usual in the genus *Hyalina*. The central teeth are small in proportion to the laterals.

Jaw already figured by Leidy (Terr. Moll. U. S. I., pl. XII. fig. 7.)

ZONITES GULARIS, Say.

Jaw highly arcuate, ends attenuated, anterior surface smooth, cutting edge with a well developed median projection.

The lingual membrane has been described by us on p. 293 of Land and Fr. W. Shells, Part I.

1-13 W. H. Dall New Mollusca of Western North America
14-16 R. E. C. Stearns New Mollusca

Conchological Memoranda, No. VII.

August 28, 1871.

Preliminary Descriptions of New Species of Marine Mollusca from the West Coast of North America.

By Robert E. C. Stearns.

Pholas Pacifica; *Stearns*.

Shell oblong, beaks two-fifths of length of shell from anterior end; anterior end of valves triangular, pointed; anterior dorsal edge of valves reflected and folded down on the umbos; lower anterior margin curved, forming a large elliptic-oval gape; posterior end of valves squarely rounded; shell dull chalky white, sculptured in concentric lines, which anteriorly are laminated and posteriorly become extinct; valves radiately ribbed, which also become obsolete at the posterior end; at the intersection of the radiating and concentric lines the sculpture is pectinated; an area below the umbos, nearly or quite destitute of sculpture, which varies much in prominence in different specimens; accessory plate sub-lanceolate and bent down on the beaks, anteriorly prolonged over but not covering the ante-umbonal gape; interior of valves white, enamelled; internal rib short, curved and flattened. Largest specimen, two and six-tenths inches in length, and one and five-tenths inches in height. Habitat—Alameda, San Francisco Bay, California, where in some places it is common in sandy mud between tide marks. Numerous specimens collected by Messrs. Harford, Hemphill, Drs. Kellogg and W. P. Gibbons.

Psychrhactus occidentalis; *Stearns*.

Shell elongated, fusiform, rather slender, whitish, traversed by narrow, revolving, brownish threads and much wider intervening spaces; suture distinct, spire tapering; aperture oblong-oval, about half the length of the shell; within white, polished; canal short, nearly straight; columellar obliquely, not strongly plicated; length about three-fourths of an inch. Habitat—near the Island of Attou, at the west end of the Aleutian Archipelago.

Fusus (Chrysodomus?) Harfordi; *Stearns*.

Shell solid, elongate, regularly fusiform; spire elevated, whorls six or seven, moderately convex, slightly flattened (in outline) above, with a groove or channel following the suture; color, chocolate brown; surface marked by numerous narrow revolving costæ, which alternate in prominence on the body whorl, and longitudinally by fine incremental striæ, and on the upper whorls by obtusely rounded ribs of more or less prominence; aperture ovate, about one-half the length of the shell, polished, white and finely ribbed within; (the outer lip in perfect specimens is probably finely crenulated); canal short, nearly straight. Lon. 2 I; Lat. .94 in. Number of Specimens, three; two mature, dead, one junior, fresh. Habitat—coast of Mendocino County, near Big Spanish Flat, California, where it was detected by Mr. Harford.

PLEUROTOMA (DRILLIA) MONTEREYENSIS: *Stearns.*

Shell small, rather solid, elongate, slender; spire elevated, sub-acute; whorls seven to eight moderately rounded; upper portion of larger volutions somewhat concavely angulated; suture distinct; color, dark purplish brown or black; surface covered with rather coarse, inconspicuous, revolving costæ, interrupted on the body whorl by rude incremental lines; middle of upper whorls and upper part of body whorl displaying fourteen to fifteen equidistant, longitudinal, nodose, slightly oblique ribs, which are whitish in the specimen before me (being somewhat rubbed) on the larger whorls; on the smaller volutions of the spire a puckering at and following the suture suggests a second indistinct series of nodules; aperture less than half the length of the shell; canal short; terminal portion of columella whitish, slightly twisted; posterior sinus, rather broad rounded, and of moderate depth. Mean divergence about 26°. Long. .67 in.; Lat. .24 in. Habitat—Monterey, California, where the single specimen in my cabinet was collected by Mr. Harford and myself in March, 1868. The shell, in its general aspect, resembles the sombre colored species of the Gulf of California and Panama.

PLEUROTOMA (DRILLIA) HEMPHILLII: *Stearns.*

Shell small, smooth, slender, polished; spire long, subacute, rounded at apex; longitudinally marked with inconspicuous, oblique ribs, which are nearly obsolete on the body whorl; number of whorls seven, with well defined sutural line, and just below it a parallel impressed thread-like line; shell of an opaque dingy horn color; incremental lines fine, marked in some specimens with dingy white; mouth obliquely ovate, about one-third the length of the shell; labrum produced, anteriorly somewhat thickened; sinu-sutural, deep, calloused; columella thickened at base; canal very short, somewhat produced and twisted; one specimen shows obscure, revolving, impressed lines below the swell of the body whorl; size quite uniform. Lon. .26; Lat. .09 inch. Habitat—Los Todos Santos Bay, Lower California, where several specimens were obtained by Mr. Hemphill, for whom I have named this well marked species.

Conchological Memoranda. No. VIII.

October 14, 1871.

DESCRIPTION OF A NEW SPECIES OF VERONICELLA FROM NICARAGUA.

BY ROBERT E. C. STEARNS.

Veronicella, Blainville. **Vaginulus**, Ferussac.

Veronicella olivacea: Stearns.

Animal elongated oval, slug-shaped, sides moderately curved, ends obtusely rounded; substance (in alcohol) coriaceous, back convex and granulously rugose; color olive beneath, darker olive above; length of body nearly four times its width; foot linear, not quite as long as, and one third the width of, the body; eye peduncles short, annulated, with rather obscure stumpy (bifurcate?) tentacles below.

Length of largest specimen 1.74 inches. Breadth of largest specimen .54 inch.

Habitat — Nicaragua (Occidental department), where several specimens were collected by Mr. J. A. McNiel.

My collection contains three specimens, and the Museum of the Peabody Academy of Science, at Salem, Mass., numerous examples of this species. In connection with the above measurements, it should be borne in mind that the contraction caused by the alcohol, materially affects the proportions; the animal, when alive, is undoubtedly very much longer, and somewhat broader, than above stated.

This species is found also in the Upper Californian province, a specimen having been collected by me near Lobitos, in the year 1866.

The few species known inhabit tropical or semitropical climates; the form above described is quite distinct from *V. Floridana*, which is also found in Nicaragua (Eastern department), where it was collected "Under stones, Javate, Chontales; probably the same species, but twice the size of Toro Rapids." Vide paper "On the Land and Fresh Water Shells of Nicaragua, by Ralph Tate"; in American Journal of Conchology, Vol. V, pp. 154-162. The "Toro Rapids" specimens of Mr. Tate's collection, possibly belong to the species herein described, but it is hardly probable that the well marked differences between the latter and *V. Floridana* could have escaped detection.

CONCHOLOGICAL MEMORANDA.

NO. IX.

[From Proceedings Cal. Acad. Nat. Sci., Sept 4, 1871.]

Mr. Stearns read the following,

On the Habitat and Distribution of the West American Species of Cypræidæ, Triviidæ and Amphiperasidæ.

Being corrections to Mr. Roberts' Catalogues * *of the "Porcellanidæ" and Amphiperasidæ.*

BY ROBERT E. C. STEARNS.

Mr. Roberts' Catalogues—if the West American species are any criterion—throw no new light upon the distribution of the species enumerated therein; neither are the more accurate statements of authors more reliable in this respect than Reeve, Sowerby and Kiener—whose indefiniteness and errors he has blindly repeated therein—referred to or considered.

The remarks of Mr. J. H. Redfield, on page 88, Volume V, of American Journal of Conchology, on the sins of the late Mr. Reeve in this respect, apply to Mr. Roberts, for it is not too much to expect, or even demand of an American writer, that he should be aware of and correct errors of the kind referred to herein, at least, when said errors occur in connection with American species. Mr. Redfield says: "The frequent errors of statement in regard to habitat are, perhaps, the most mischievous fault that can be brought against the work, for on them are liable to be based erroneous conclusions in regard to the important questions of geographical distribution, and of permanence in species."

If the species, the habitat of which is corrected by me in this paper, were from remote localities or little known, it would be less cause of surprise; but the Check-Lists of the Smithsonian Institution, Carpenter's Reports to the British Association, Dr. Cooper's Geographical Catalogue of the Mollusca, published in connection with the work of the Geological Survey of California, and sundry local lists of my own in the PROCEEDINGS OF THE CALIFORNIA ACADEMY OF NATURAL SCIENCES, were equally as accessible to Mr. Roberts as the monographs from which the catalogues were compiled.

---

* The Catalogues referred to were published in connection with the "Am. Jour. of Conch.," Vol. V, Part III.

If species appeared in the lists without habitat, it would be far preferable to positive error. Mr. Roberts' catalogues are not alone open to criticism; for others that have appeared in connection with the Journal contain omissions and geographical inaccuracies which might have been avoided by the furnishing of proofs to those investigators who, from residence or special study, possess the requisite data. The criticism herein is not dictated by hypercritical or unkind feeling, but prompted solely by a high regard for the cause in which we are all working—with, let us hope, some degree of usefulness.

1. Luponia albuginosa, *Mawe*, "California," should be *Lower* California. Although this species is credited to the Oregonian and Californian Province in the Smithsonian Institution Check-List, by Dr. Carpenter, (June, 1860), he properly omits it in his Supplemental Report to the British Association, 1863. It is common at Cape St. Lucas and various points in the Gulf of California, and belongs to the "Mexican and Panamic province."

32. L. Goodalii, *Gray*, is credited to Lord Hood's Island, which is quite indefinite, as many islands have been so named—one in the Gallapagos group, which would connect this species with the Central American fauna. This species, however, is not found there, but pertains to the Indo-Pacific province—perhaps to "Lord Hood's Island" in the Paumotu group.

49. L. onyx, *Linn.*, "San Diego I." If, by the habitat given, San Diego or the islands off that coast are meant, it may be intended for one of the Coronados; but *L. onyx* is not found there—neither at San Diego or at any other point on the west coast of America. As the species in some of its varieties resembles in coloration 69, the blunder on the part of the original author may be thus explained.

68. L. Sowerbyii, *Kiener*, for which no habitat is given, is found in the Gulf of California, and consequently belongs in the same province with No. 1. Some authors have confounded this species with L. picta, *Gray*, which latter is African. In Sby's Conch. Illust., L. picta is credited to "Guaymas;" while L. Sowerbyii, through its synonym, "zonata, *Lam*," is without habitat.

69. L. spadicea, *Swains.*, "New Holland," is a Californian, credited by Dr. Cooper to "Santa Barbara, San Diego and islands," which is correct, being confirmed by my published and manuscript lists. It is a well-marked species and quite distinct from 49. It is figured in "Chenu's Manuel," Vol. I, fig. 1715—the outline of which is well enough, but the dark spots so prominently represented might lead astray.

4. Aricia arabicula, *Lam.*, is properly credited to "Acapulco," though its specific centre is in the neighborhood of Mazatlan, Gulf of California, where it is quite common. Its occurrence at Panama is exceedingly rare. Professor C. B. Adams found but "7 specimens on the reef"; while of fifty-nine (Luponia) punctulata, *Gray*, credited in the List to "Mazatlan-Panama," he collected "335 specimens." Mazatlan is quite likely an error, for in the great mass of material from that place, and from other points on the Gulf of California which has passed under my examination, I have never detected a specimen. It is sometimes washed up dead by the winter storms at Cape St. Lucas.

We find in the Genus Trivia of Mr. Roberts' list, numbered "10, T. depauperata, *Sowb.*, California," which gives us one species more than we claim; unfortunate? specimens of "5. T. Californica, *Gray*, California," after receiving hard treatment in the surf and gravel on the beach—where Mr. Sowerby's specimens were, without doubt, obtained—were finally recompensed by specific honors. Having collected numbers of specimens at various points along the coast, all of which have been carefully examined and compared, I have no hesitation in placing Mr. Sowerby's species as a synonyme of T. Californica, which extends southerly to the Gulf of California, where specimens are occasionally detected.

23. Trivia Pacifica, *Gray*. "Gallapagos," is correct; but the species is also found in the Gulf of California, and it is from the latter place that most of the specimens in cabinets have been obtained; it is not a common shell, and from its resemblance to

39. T. suffusa has been confounded with the latter, and the latter has, in some instances, been wrongly credited to the Gallapagos. T. suffusa appears to be the W. Indian analogue of T. Pacifica.

29. T. pulla, *Gask.*, "Gallapagos Islands;" quite likely correct; it is a rare species; Carpenter reports one specimen in the Mazatlan collection of Reigen and I have it in my cabinet from the Gulf of California.

31. T. radians, *Lam.*, "Mazatlan-Ecuador;" these localities are correct, but it can justly claim a more northern extension than Mazatlan, as it was collected by Mr. Gabb, in 1867, at San Juan, Lower California, and Dr. Cooper probably collected it at San Pedro, in the Oregonian and Californian province, as he credits it to that place, which is about a *thousand miles* north of Mazatlan. It is said to extend south of the equator to Peru, from which place, I think, I have received specimens. It belongs to the W. Mexican and Panamic province.

34. T. sanguinea, *Gray* "Mazatlan-Ecuador." As to the southern limit of this species I have no data; but its northern line must be moved about the same distance as the previous species, viz.: to Catalina Island, off the coast of California, and it belongs in the same province with 31.

36. T. Solandri. *Gray*, "Pacific Ocean;" from which it would be inferred that this species was an Indo Pacific, rather than a West American form. Its specific centre is the Gulf of California, where it is quite abundant; it has been collected on the coast of California by Dr. Newcomb, also by the late Mr. Hepburn, as far north as Santa Barbara, * and Dr. Cooper credits it also to San Nicholas Island. It has not been reported south of Acapulco, and belongs to the northern part of the Mexican and Panamic province.

38. T. subrostrata, *Gray*, "West Indies." Undoubtedly correct. In the Mazatlan catalogue (species 444), Dr. Carpenter mentions a *Trivia* closely resembling this West Indian form, which, on the strength of Dr. Gray's identi-

---

* See my lists of Hepburn's and Newcomb's collections at Santa Barbara, etc., in the Pro. Cal. Acad. Nat. Sci., Vol. III, pp. 283–286, and pp. 343–345.

fication, was listed under the above name. That the solitary Mazatlan shell referred to is identical with the West Indian, is highly improbable.

The highest northern station on the west coast of America at which any representative of the Porcellanidæ (Cypræidæ) has been detected, is the rocky point known as Bodega Head, some fifty miles north of the entrance to San Francisco Bay, in latitude about 38° north. This is one hundred and forty miles farther north * than the species (Trivia Californica or any other related form) has been reported prior to my Bodega collection in June 1867.

Passing to the Amphiperasidæ—the catalogue of which is published in connection with the Porcellanidæ—it will not be irrelevant to direct attention to the paper of Professor Gill "On the Relations of the Amphiperasidæ," published in the Am. Jour. of Conch., Vol. VI. pp. 183-187. The marked difference in the form and plan of structure of the shells of this family, with the exception of A. ovum (which, in a general way, resembles Cypræa) as well as the anatomical differences indicated by Professor Gill, require that the forms included in the catalogue referred to should be removed from a consecutive classification. That they more nearly approach the form known as Pedicularia is readily seen. As the shells of the latter are rare in collections, and are quite important to the student in this connection, I would suggest an examination of the red and purple corals of the Indo-Pacific waters, upon which, by careful scrutiny, specimens may frequently be found, of the same color as the coral to which they are attached.

7. Volva avena, Sowb., "Santa Barbara–Panama," has never been confirmed so far north as Santa Barbara.

36. V. variabilis, C. B. Ad., " Cape St. Lucas," has a more northern limit, having been collected at San Pedro, California, by Dr. Cooper, and Carpenter credits it to Santa Barbara, ("Jewett.") I am inclined to believe that V. avena = V. neglecta of C. B. Ad., Mr. Sowerby's name having priority by twenty years.

10. V. Californica, Sowb., MSS., "California," has never been confirmed from any point within the Oregonian and Californian Province, and is undoubtedly an error.

31. V. similis, Sowb., for which no habitat is specified, should be credited to the Gulf of California.

All of the West American species are well represented in my collection, though, with the exception of V. variabilis they may be justly considered as rare.

---

* I refer to American species.

DESCRIPTIONS OF NEW SPECIES OF MARINE MOLLUSKS FROM THE COAST OF FLORIDA. BY ROBERT E. C. STEARNS.

Marginella (Glabella) opalina Stearns.

Shell ovate, solid; light to dark amber, some specimens showing obscure bands, more or less intense, of same color; subtransparent, smooth, polished; spire elevated, apex rounded; whorls four, suture distinct; aperture rather more than half the length of the shell; outer lip thickened, internally crenated, and strongly notched above; columella with four well developed plaits.

Largest specimen measured long. .21, lat. .1 inch.
Smallest do.   do.   long. .17, lat. .09 inch.

Habitat: Rocky Point, Tampa Bay, west coast of Florida, where several specimens were collected by Col. E. Jewett and myself; this beautiful little shell was found by us upon the under side of bunches of oyster shells, near low water mark. I know of no other species with which it might be confounded.

Marginella (Glabella) auræcincta Stearns.

Shell small, solid, ovate-conic; spire elevated, rounded at the apex; whorls, five, suture indistinct, being hidden by the enamel; aperture narrow, linear, about half the length of the shell; outer lip thickened, its internal edge moderately notched above and crenated below; surface smooth, polished, white, with two revolving amber-colored bands; columella with four prominent plications.

Measurement: long., .16, lat., .07 inch.

Habitat: Long Key, on the west coast of Florida, where I obtained the single (living) specimen described. An exceedingly beautiful shell, resembling in its general features, color excepted, my *G. opalina*, but less robust, with a more acute spire; the internal crenations of the outer lip less prominent; the columellar plaits less conspicuous, closer and more oblique. The color of the bands will quite likely be found to vary in different individuals; in my solitary specimen the bands are a light, clear amber, golden when seen through the intensified light of a magnifier, suggesting the gilded striping on French porcelain.

Drillia ostrearum Stearns.

Shell small, elongated, slender; spire elevated, subacute; whorls, seven or eight, concavely angulated above and moderately convex below; longitudinally sculptured with (16-20) rounded ribs, inconspicuous on the angle, most prominent upon the extreme convexity of the whorls, decreasing and becoming obsolete anteriorly; intersections of sculpture nodulous; suture marked with a thread-like, revolving rib; color, dingy yellow to purplish black; aperture ovate, narrow, about two-fifths the length of the shell; outer lip thin, with a rounded, shallow notch near the suture; columella nearly straight, canal short.

Measurement: long., .67, lat., .24 inch, largest specimen; long., .51, lat., .18 inch, smallest specimen; number of specimens, three.

Habitat: Pine Key, Tampa Bay, west coast of Florida, where I found them on bunches of oysters, in a shallow pool overflowed by the tides. The sculpture varies in prominence in different individuals.

Mangelia stellata Stearns.

Shell small, fusiform, turrited, yellowish tinged, more or less with reddish brown; number of whorls seven, angulated above; suture distinct; sculptured with twelve to thirteen strong, smooth, longitudinal ribs, which extend to the extremity of the basal volution, which also shows near its termination a few revolving lines; intercostal spaces marked by fine incremental striæ; aperture narrow, rather oblique, less than half the length of the shell; labrum effuse, externally much thickened, and deeply notched near the suture; canal short; columella somewhat callous and bent forward.

Measurement: length of largest specimen, .35 inch, lat., .14 inch.

Habitat: Tampa Bay, west coast of Florida, where I obtained specimens on the mainland at Rocky Point, and in a lagoon at Point Penallis, upon an oyster bar in lagoon at Pine Key, also on Long Key.

This well marked species varies in color from light buff to dark ferruginous brown; the mouth in most of the specimens is of the latter color. Viewed from above this simple shell has the form of a many short-pointed star.

Architectonica tricarinata Stearns.

Shell small, solid, trochiform, moderately elevated; whorls four, angulated, with three equidistant, prominent, revolving ribs on the periphery of the basal volution and two on the whorl above; suture distinct, sometimes marked with an inconspicuous, submodose rib; aperture round; peritreme much thickened; umbilicus profound, strongly crenulated; color of shell white, more or less spotted and blotched with light red or dark umber; number of specimens, seven.

Measurement. The largest measures long., .12 inch, lat., .16 inch.

Habitat: Long Key and shores of mainland, Tampa Bay, west coast of Florida, where I found it (dead) on the beaches; it was also collected by Col. E. Jewett and Dr. Wm. Stimpson.

Specimens vary one from another in sculpture and color markings; in one specimen the sutural rib is of equal prominence with the other ribs, and in all the specimens the sutural rib is broken into slight nodules, which are, as well as the umbilical crenulations, regularly spotted with dark umber; other specimens show inconspicuous riblets on the external portion of the outer lip, which soon blend into the general surface of the shell.

Siphonaria naufragum Stearns.

Shell oval, depressed conic, with numerous fine radiating ribs of a whitish color, the interspaces of a reddish or chocolate brown; also marked by many fine and occasional, irregular, coarser, concentric striæ; edge of the shell internally finely crenulated; muscular impression distinct; siphonal groove shallow, inconspicuous; shell moderately convex in outline; apex recurved, subcentral, nearer the posterior margin; interior of shell enamelled, with same color marks as externally; number of specimens, twenty-five.

Measurements of largest and smallest as follows: long., .93 inch, lat., .71 inch, alt., .47 inch: long., .73 inch, lat., .54 inch, alt., .35 inch.

Habitat: Outer beach of Amelia Island, east coast of Florida, upon the timbers of an old wreck,[1] near low water mark; numerous specimens of this fine species were collected at the same time by my friends Col. E. Jewett and Dr. William Stimpson.

Carithidea turrita Stearns.

Shell small, elongately conic, rather delicate, purplish white to dark purple, with whitish revolving band on the middle of the whorls, inconspicuous except in the aperture; spire gradually tapering; whorls twelve, moderately convex, with sixteen to twenty prominent, smooth, equidistant, whitish longitudinal ribs, which terminate abruptly a little below the periphery of the last whorl, with a single narrow, revolving keel below; suture deeply grooved; anterior portion of bodywhorl smooth or marked only by incremental lines; aperture rounded above, subquadrate below; outer lip effuse, externally thickened; labium anteriorly prolonged, angulated; in some specimens the peristome is continuous. Number of specimens (adult) about one hundred, varying in measurement from lon., .51, lat. .16 inch to lon. .33, lat. .11 inch.

Habitat: Point Penallis, Tampa Bay, west coast of Florida, where I found it abundant beneath a confervoid growth in a shallow lagoon, associated with *Cyrena Floridana*; also Mullet Key. Col. E. Jewett; Dr. Stimpson found specimens at other points on Tampa Bay.

C. turrita is much smaller than Say's species (*C. scalariformis*), which has a greater number of longitudinal and several revolving ribs; it is a more delicate and handsomer shell than *C. ambiguum* C. B. Ad., from Jamaica, which it somewhat resembles; it is smaller than *C. costata*, of Da Costa, from New Providence in the Bahamas, which latter is more finely and closely ribbed. All of the above species possess characteristics in common, which place them naturally in the same group, analagous to the West American group, of which *C. Montagnei* and *C. pulchrum* are representatives.

[1] See "Rambles in Florida," in Am. Naturalist, vol. iii, p. 287.

CONCHOLOGICAL MEMORANDA.

No. X.

[From the Proceedings of the California Academy of Sciences, June 3, 1872.]

Description of a New Species of Mangelia, from California.

BY ROBERT E. C. STEARNS.

Mangelia interlirata, Stearns. Shell of a dark reddish brown, small, solid, slender, fusiform; whorls, eight, prominently sculptured with 8-10 strong longitudinal and 10-12 thread-like revolving ribs, the latter of a darker shade and meeting, but not crossing the former; aperture linear, less than half the length of the shell; outer lip simple, somewhat thickened, externally and posteriorly, slightly notched; number of specimens, four, of which two are measurably perfect. The dimensions of the largest is

Long. ·27; Lat. ·09 inch.

Habitat, Monterey, California, where the above specimens were found dead on the beach by Mr. Harford and myself.

A larger and more perfect specimen than either of the four above mentioned, also from Monterey, was given by me to Dr. Cooper, in 1865, for the State Museum, but does not appear to have been described by him. I think that the same form, which is quite rare, has been detected at San Diego by Mr. Hemphill, but am not positive. It is with some diffidence that I place it in *Mangelia* although it seems to accord exceedingly well with the typical figure *M. striola ta*, of Schacchi. (Vide, Adams's Gen. pl. 10, fig. 10*a*.)

[From the Proceedings of the California Academy of Sciences, July, 1, 1872.]

Remarks on Marine Faunal Provinces on West Coast of America.

BY ROBT. E. C. STEARNS.

Mr. Stearns called the attention of the Academy to certain provincial divisions in the marine faunæ of the West Coast of America, suggested by Prof. A. E. Verrill, in the transactions of the Connecticut Academy for 1871; and remarked more particularly on that part of the coast from Cape St. Lucas, northward, that to divide this portion upon the data at present known, so as to make provinces which shall correspond with those of the Atlantic side, is not warranted by the knowledge possessed at the present time; that the topography and geology of the portion of the West American Coast, specified by him, was much more uniform in its character, as well as in the temperature of its waters, than a corresponding extent of the Atlantic coast, to say nothing of the influence of the coast currents which upon this side are peculiar and enter largely in the matter of distribution of species; furthermore, that the MSS. data in his possession which were, to say the least, fully as important as what had already been published, and quite likely more authentic, indicated a greater range for each province, and therefore a less number of provinces than suggested by Prof. Verrill.

Though much had been done by himself and other members of the Academy co-operating with him in the accumulation of data bearing upon the geographical distribution of the mollusca of our coast, still so much remained to be done, in order to make the work thorough and reliable, that it would be merely arbitrary, and necessarily requiring frequent readjustment, to propose at this time any new divisions or subdivisions of the coast into zoological provinces.

As to that part of the West Coast of North America from

Cape St. Lucas, including the Gulf of California, thence southerly to a point a few miles south of Panama, with the exception of collections made at a few places in the Gulf of California, also at San Juan del Sur, and its immediate vicinity on the coast of Nicaragua and in the bay of Panama, but little more is known of this vast reach of shore-line, than was known years ago.

Mr. Stearns stated that at some future time, as soon as the data collected by himself and his co-workers here are compiled, he proposed to refer to the subject again.

[From the Proceedings of the California Academy of Sciences, August 5th, 1872.]

Description of New Species of Shells from California.

BY ROBERT E. C. STEARNS.

Siphonaria Brannani, Stearns. Shell oval, subconical, helcion-shaped; apex recurved and somewhat twisted, anterior and sometimes quite in line with margin; surface of shell irregularly undulating, of a dark brownish color, and marked with numerous fine whitish radiating ribs which crenulate the margin; shell internally shining, and dark chocolate brown; muscular impression and siphonal groove distinct. Some specimens are quite irregular in outline, being affected in that respect by the inequalities of the surface upon which they are found. Numerous specimens of this shell were collected at Santa Barbara Island, off the southern coast of this State, in the month of June, 1871, by Mr. S. A. L. Brannan, to whom I am indebted for the specimens from which this description is made. The largest of eighteen specimens measures, long. ·39, lat. ·30 inch, though most of the specimens are much smaller than above dimensions.

Truncatella Stimpsoni, Stearns. Shell cylindrical, solid, light reddish horn-color, or amber; shining, slightly decreasing in size towards apex; closely and strongly longitudinally ribbed, the ribs even, regular and interrupted only by the suture; upper whorls wanting, remaining whorls, 4; aperture oval, somewhat oblique, slightly angulated above; peristome continuous, thickened and moderately angulated at its junction with the body whorl.

Length of largest specimen, ·22 inch; length of aperture, ·06 inch.

Habitat; False Bay, near San Diego, California, where numerous specimens were detected by Henry Hemphill, Esq. This shell is quite distinct from *T.*

Californica Pfr, the latter having an almost smooth surface. Specimens of *T. Stimpsonii* are in the cabinets of Messrs. Henry Hemphill, W. G. Binney, and Thomas Bland, the Philadelphia Academy of Natural Sciences and my own.

[From the Proceedings of the California Academy of Sciences, August 19, 1872.]

Notes on Purpura canaliculata, of Duclos.

BY ROBERT E. C. STEARNS.

This fine species ranges from Unalashka, south to Monterey, California—most of the specimens heretofore distributed being from the intermediate point of Vancouver Island. Specimens from the last named place, large numbers of which I have examined, are less variable in size and form than those from farther north, and are generally a more delicate shell. One marked peculiarity of the Vancouver specimens is the wide groove or sulcation following the suture, which appears to be constant, as I have found it in all the specimens from this locality; it is infrequent in the more northern and southern specimens. The Vancouver shells average $1\tfrac{1}{8}$ inches in length, and the costæ are very prominent. The specimens from Unalashka are the largest I have seen, the average length of a large number (60) reaching $1\tfrac{5}{6}$ inches, and a few specimens measuring $1\tfrac{7}{8}$ inch. A variety from Sitka is of a dingy yellow color, (adults) internally of a brownish yellow, sometimes running into a dark reddish brown; occasionally a banded variety is met with at all of the localities. The shells of this species are generally prolonged anteriorly, which gives a somewhat acute V-shape to that part of the mouth.

Comparing the specimens, from Unalashka to Monterey, including intermediate places, it will be seen that they vary greatly in size, color and general outline, as do the other species of Purpuridæ found within the same limits. The general variation from the Vancouver form is in the much larger and ventricose body-whorl.

[From the Proceedings of the California Academy of Sciences, October 7, 1872.]

A partial comparison of the Conchology of portions of the Atlantic and Pacific coasts of North America.

BY ROBERT E. C. STEARNS.

A striking feature in the Conchological fauna of that part of the Pacific coast included in the Californian and Vancouver Zoological province when compared with the molluscan fauna of the Atlantic coast from the Arctic seas to Georgia, is the preponderance in the former of those forms of molluscan life which are embraced in the Order of Scutibranchiata.*

The Scutibranchiate Gasteropods, or shield-gilled crawlers, comprise a great number of mollusks, all of which are marine, and which inhabit the sea shore principally the littoral and laminarian zones, subsisting on marine vegetation; thus we find the beautiful group of *Callostoma* upon the larger algæ as well as the unique *Trochiscus* (*T. Sowerbyi*), and *Chlorostoma* crawling over the sedimentary rocks, upon which grows the green Cladophora or some allied vegetable form upon which it feeds, and which also is the favorite food of several species of limpets.

The Order of Scutibranchiata according to the Adams's, includes the families of *Neritidæ* (none of which are found in the Californian and Oregonian province, though they begin to appear on the coast of Lower California); the *Trochidæ*, which is largely represented by the following genera: *Eutropia* one species; *Leptothyra* three species; *Pachypoma* and *Pomaulax* one species each; *Liotia* one, perhaps two species; *Thalotia* and *Trochiscus* one species each; *Callistoma*, *Chlorostoma*, *Omphalius*, *Margarita* and *Gibbula* each by several species.

The Family of *Haliotidæ* which is represented by several species all of large size, widely distributed and exceedingly numerous in individuals; *Fissurella* including *Lucapina*, *Glyphis* and *Clypidella*, also *Puncturella* and *Emarginula*.

Dentaliidæ by two or more species; *Tecturidæ* by several species of *Acmæa* also by *Scurria*; *Gadinia* by one and *Nacella* by six or more species.

Chitonidæ by numerous species and great numbers of individuals.

It may be that some of the groups included by the Messrs. Adams in the Order referred to, as our knowledge increases, will require to be separated or removed, but so far as the purposes of comparison as made herein are considered, the result will not be materially impaired.

The total number of marine molluscan species and well marked varieties within the Californian and Oregonian province, so far as known and determined, is not far from 630, of which about 200 are Bivalves; and of the remaining 430, 123

---

*Vide Adams' Genera of Recent Mollusca, Vol. I. p. 376.

are included within the Scutibranchs; of this latter number about 40 belong to the Chitonidæ and the same number to the Trochidæ.

Of the 247 marine gasteropods enumerated by the late Dr. Stimpson in the Smithsonian Institution Check-list, as found from the Arctic Seas to Georgia, 32 only, or less than one-eighth, come within the Order mentioned; of this comparatively small number seven (7) are *Chitons* and fourteen (14) belong to the *Trochidæ*, while *Haliotis*† is without a representative; the *Trochidæ* within this province are not represented by such marked or unique characters as distinguish their relatives on the West Coast.

Some revision may be required hereafter in the number of Scutibranchiate species credited to the West coast province, as forms now catalogued as distinct may in some instances be united; but on the other hand, it is not unlikely that new forms undoubtedly distinct will be detected when the coast is more thoroughly explored.

† A solitary specimen of *Haliotis*, of small size, was obtained through dredging in the Gulf Stream, four or five years ago, by Count L. F. Pourtales, of the U. S. Coast Survey, but south of Georgia.

INDEX TO PLATES, VOL. IV.

PLATE I.

		PAGE
Fig. 1.	*Voluta (Scaphella) Stearnsii*, Dall, $\frac{1}{1}$	270
Fig. 2.	*Nacella* (?) *rosea*, Dall. $\frac{2}{1}$	270
Fig. 3, 3a.	*Littorina Aleutica*, Dall, $\frac{2}{1}$	271
Fig. 4, 4a.	*Siphonaria Brannani*, Stearns, $\frac{2}{1}$	249
Fig. 5.	*Truncatella Stimpsonii*, Stearns, $\frac{2}{1}$	248
Fig. 6.	*Magasella Aleutica*, Dall, $\frac{2}{1}$	302
Fig. 7.	*Terebratella occidentalis*, Dall, $\frac{2}{1}$	182
Fig. 8.	*Amphissa* (? *versicolor*, Dall, *var.*) *lineata*, Stearns, $\frac{2}{1}$.	—
Fig. 9*.	*Pedicularia Californica*, Newc., (Vol. III, p. 121) $\frac{2}{1}$.	—
Fig. 10.	*Mangelia interlirata*, Stearns, $\frac{2}{1}$	226

* *Pedicularia Californica*, described by Dr. W. Newcomb, on page 121 of Volume III of this Academy's Proceedings, (now figured from a specimen in my collection,) has again been detected on Corals, in deep water near the Farallones; also by G. W. Dunn and the late Dr. C. A. Canfield, at Monterey.—R. E. C. Stearns.

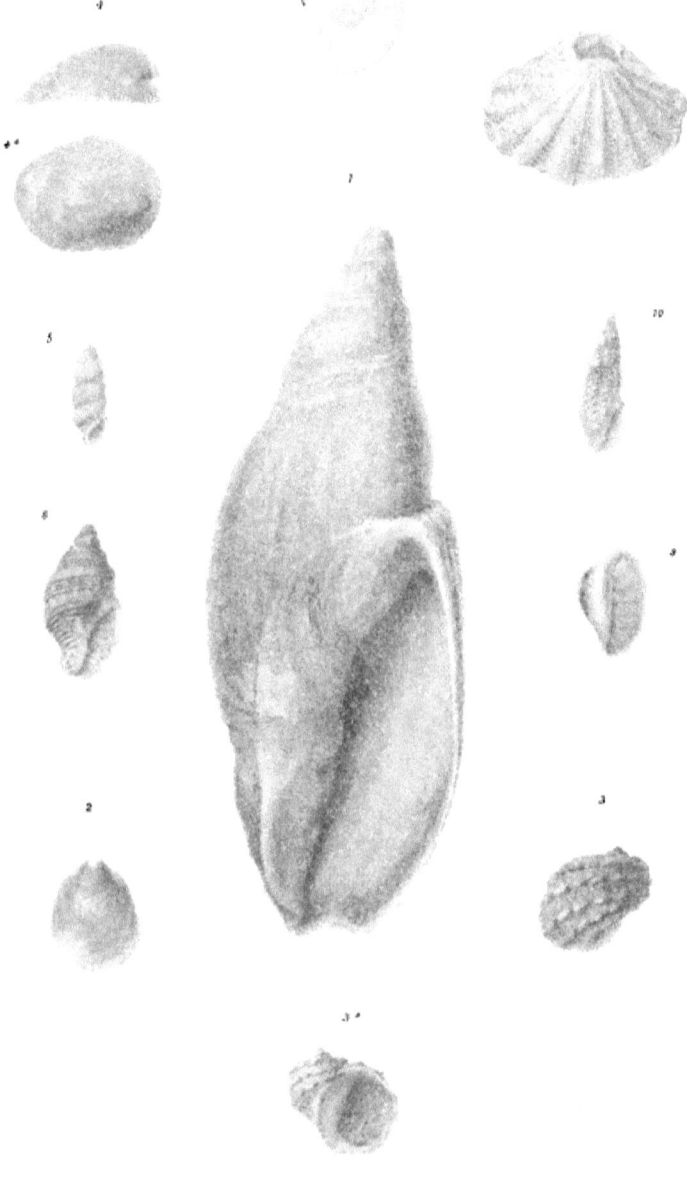

CONCHOLOGICAL MEMORANDA.

No. XI.

[From the Proceedings of the Boston Society of Natural History, January 17, 1872.]

Descriptions of New Species of Marine Mollusks from the Coast of Florida.

BY ROBERT E. C. STEARNS.

Marginella (Glabella) opalina, Stearns.

Shell ovate, solid; light to dark amber, some specimens showing obscure bands, more or less intense, of same color; subtransparent, smooth, polished; spire elevated, apex rounded; whorls four, suture distinct; aperture rather more than half the length of the shell; outer lip thickened, internally crenated, and strongly notched above; columella with four well developed plaits.

Largest specimen measured long. .21, lat. .1 inch;
Smallest specimen measured long. .17, lat. .09 inch.

Habitat: Rocky Point, Tampa Bay, west coast of Florida, where several specimens were collected by Col. E. Jewett and myself; this beautiful little shell was found by us upon the under side of bunches of oyster shells, near low water mark. I know of no other species with which it might be confounded.

Marginella (Glabella) aureocincta, Stearns.

Shell small, solid, ovate-conic; spire elevated, rounded at the apex; whorls, five, suture indistinct, being hidden by the enamel; aperture narrow, linear, about half the length of the shell; outer lip thickened, its internal edge moderately notched above and crenated below; surface smooth, polished, white, with two revolving amber-colored bands; columella with four prominent plications.

Measurement: long. .16, lat. .07 inch.

Habitat: Long Key, on the west coast of Florida, where I obtained the single (living) specimen described. An exceedingly beautiful shell, resembling in its general features, color excepted, my *G. opalina*, but less robust, with a more acute spire; the internal crenations of the outer lip less prominent; the columellar plaits less conspicuous, closer and more oblique. The color of the bands will quite likely be found to vary in different individuals; in my solitary specimen the bands are a light, clear amber, golden when seen through the intensified light of a magnifier, suggesting the gilded striping on French porcelain.

Drillia ostrearum, Stearns.

Shell small, elongated, slender; spire elevated, sub-acute; whorls, seven or eight, concavely angulated above, and moderately convex below; longitudinally sculptured with (16–20) rounded ribs, inconspicuous on the angle, most prominent upon the extreme convexity of the whorls, decreasing and becoming obsolete anteriorly; intersections of sculpture nodulous; suture marked with a thread-like, revolving rib; color, dingy yellow to purplish black; aperture ovate, narrow, about two-fifths of the length of the shell; outer lip thin, with a rounded, shallow notch near the suture; columella nearly straight, canal short.

Measurement: long. .67, lat. .24 inch, largest specimen; long. .51, lat. .18 inch, smallest specimen; number of specimens, three.

Habitat: Pine Key, Tampa Bay, west coast of Florida, where I found them on bunches of oysters, in a shallow pool overflowed by the tides. The sculpture varies in prominence in different individuals.

Mangelia stellata, Stearns. p. ??

Shell small, fusiform, turrited, yellowish tinged more or less with reddish brown; number of whorls seven, angulated above; suture distinct; sculptured with twelve or thirteen strong, smooth, longitudinal ribs, which extend to the extremity of the basal volution, which also shows near its termination a few revolving lines; intercostal spaces marked by fine incremental striæ; aperture narrow, rather oblique, less than half the length of the shell; labrum effuse, externally much thickened, and deeply notched near the suture; canal short; columella somewhat callosed and bent forward.

Measurement: Length of largest specimen, .35, lat. .14 inch.

Habitat: Tampa Bay, west coast of Florida, where I obtained specimens on the mainland at Rocky Point, and in a lagoon at Point Penallis, upon an oyster bar in a lagoon at Pine Key, also on Long Key.

This well marked species varies in color from light buff to dark ferruginous brown; the mouth in most of the specimens is of the latter color. Viewed from above this simple shell has the form of a many short-pointed star.

*Architectonica tricarinata, Stearns. p. ??

Shell small, solid, trochiform, moderately elevated; whorls, four, angulated, with three equidistant, prominent, revolving ribs on the periphery of the basal volution, and two on the whorl above; suture distinct, sometimes marked with an inconspicuous, subnodose rib; aperture round, peritreme much thickened; umbilicus profound, strongly crenulated; color of shell white, more or less spotted and blotched with light red or dark umber; number of specimens, seven.

Measurement: the largest measures long. .12 inch, lat. .16 inch.

Habitat: Long Key and shores of mainland Tampa Bay, west coast of Florida, where I found it (dead) on the beaches; it was also collected by Col. E. Jewett and Dr. Wm. Stimpson.

Specimens vary one from another in sculpture and color markings; in one specimen the sutural rib is of equal prominence with the other ribs, and in all the specimens the sutural rib is broken into slight nodules, which are, as well as the umbilical crenulations, regularly spotted with dark umber; other specimens show inconspicuous riblets on the external portion of the outer lip, which soon blend into the general surface of the shell.

---

* Dr. P. P. Carpenter suggests that this is a "Miuolia."

Siphonaria naufragum, Stearns, p. 23.

Shell oval, depressed conic, with numerous fine radiating ribs of a whitish color, the interspaces of a reddish or chocolate brown; also marked by many fine and occasional, irregular, coarser, concentric striæ; edge of the shell internally finely crenulated; muscular impression distinct; siphonal groove shallow, inconspicuous; shell moderately convex in outline; apex recurved, subcentral, nearer the posterior margin; interior of shell enamelled, with same color marks as externally; number of specimens, twenty-five.

Measurements of the largest and smallest as follows: long. .93 inch, lat. .71 inch, alt. .47 inch; long. .73 inch, lat. .54 inch, alt. .35 inch.

Habitat: Outer beach of Amelia Island, east coast of Florida, upon the timbers of an old wreck,* near low water mark; numerous specimens of this fine species were collected at the same time by my friends Col. E. Jewett and Dr. William Stimpson.

Cerithidea turrita, Stearns, p. 24.

Shell small, elongately conic, rather delicate, purplish white to dark purple, with whitish revolving band on the middle of the whorls, inconspicuous except in the aperture; spire gradually tapering; whorls twelve, moderately convex, with sixteen to twenty prominent, smooth, equidistant, whitish longitudinal ribs, which terminate abruptly a little below the periphery of the last whorl, with a single narrow, revolving keel below; suture deeply grooved; anterior portion of body-whorl smooth or marked only by incremental lines; aperture rounded above, subquadrate below; outer lip effuse, externally thickened; labium anteriorly prolonged, angulated; in some specimens the peristome is continuous. Number of specimens (adult) about one hundred, varying in measurement from long. .51, lat. .16 inch to long. .33, lat. .11 inch.

Habitat: Point Pennllis, Tampa Bay, west coast of Florida, where I found it abundant beneath a confervoid growth in a shallow lagoon, associated with *Cyrena Floridana*; also Mullet Key, Col. E. Jewett; Dr. Stimpson found specimens at other points on Tampa Bay.

C. turrita is much smaller than Say's species (*C. sca'ariformis*), which has a greater number of longitudinal and several revolving ribs; it is a more delicate and handsomer shell than *C. ambiguum* C. B. Ad., from Jamaica, which it resembles; it is smaller than *C. costata*, of Da Costa, from New Providence in the Bahamas, which latter is more finely and closely ribbed. All of the above species possess characteristics in common, which place them naturally in the same group, analagous to the West American group, of which *C. Montagui* and *C. pulchrum* are representatives.

---

* See "Rambles in Florida," in Am. Naturalist, Vol. III, page 267.

CONCHOLOGICAL MEMORANDA.

No. XII.

Descriptions of a New Genus and two New Species of Nudibranchiate Mollusks from the Coast of California.

BY ROBERT E. C. STEARNS.

Genus LATERIBRANCHILÆA, Stearns.

Animal like *Triopa*, with a single series of gills on each side, central or subcentral and opposite.

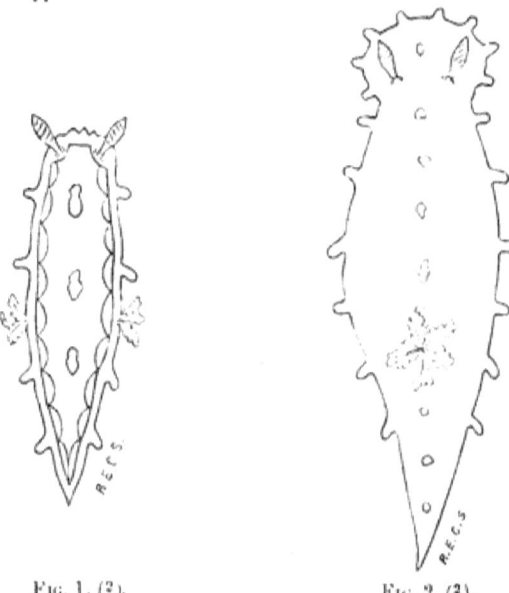

Fig. 1. (2/1).   Fig. 2, (2/1).

LATERIBRANCHILÆA FESTIVA, Stearns, Fig. 1.

Body slug-shaped, about one inch long ; of a *translucent* cream white color on back, ornamented with looped linear markings on each side, of an opaque *chalky white*, and three irregular, ring-shaped markings of the same color, nearly equidistant and along a central line on the back, also marked with a few inconspicuous irregularly placed orange spots ; cephalic tentacles short, clavate, stumpy, fringed at base, branchial orifices on each side, sub-central, with short arborescent plumes.

Habitat.—Point Pinos, near light house, Monterey, California, on the under side of granite boulders at extreme low tide; detected by Mr. Harford and myself in March, 1868.

TRIOPIDÆ, Gray.

TRIOPA, Johnston.

TRIOPA CARPENTERI, Stearns, FIG. 2.

Animal slug-shaped; anteriorly obtusely rounded, posteriorly pointed, somewhat attenuated; cephalic tentacles clavate, upper part of same of an orange color, below white; gill plumes five, arborescent, resembling fern leaves, tipped with orange; plumes and tentacles 1-16 inch in length; the former situated in middle of the back somewhat posterior to centre. Six tentacular processes on each side, tipped with orange and 1-32 inch long; also short tentacular processes in front of the head; body one and one-half inches in length, translucent white, covered with fine papillæ of an orange color.

Habitat.—Monterey, at Point Pinos near the light house, on the under side of granite rocks at edge of laminarian zone, where the above was collected by Mr. W. G. W. Harford and myself in March, 1868.

This species is named for my friend Dr. P. P. Carpenter of Montreal, whose thorough work in connection with the mollusca of W. North America has been of great service to investigators.

The above descriptions, though somewhat meagre from lack of the proper instruments for more careful diagnosis, are nevertheless adequate to a ready determination of both of the above well marked and elegant species.

Descriptions of New Marine Mollusks from the West Coast of North America.

BY ROBERT E. C. STEARNS.

CONUS DALLI, Stearns. Plate I. fig. 1.

Shell conical, robust with a smooth surface faintly marked with incremental lines; lower third portion of shell obscurely spirally ribbed and the spire elevated and indistinctly grooved on the top of each whorl; body whorl and spire moderately convex, the latter with a distinct sutural line and a faint sulcation parallel to the same; outer lip simple, aperture linear, internally of a delicate rose-pink tinge; surface of shell marked with irregular longitudinal stripes of reddish brown and sienna yellow, the former color predominating and blending in more or less and glazing the yellow; the longitudinal markings are interrupted by a series of four revolving bands (of which the two lowest are the widest.) composed of numerous whitish spots of irregular size and shape but generally small, rounded or angular; occasionally whitish subangulate spots of larger size than those included in the bands occur between the same, and in line with the longitudinal markings.

Dimensions of largest: Long. 2.35; lat. 1.22 inches. Another specimen measures: Long. 2.15; lat. 1.1 inches.

Habitat.—Gulf of California, from whence specimens are occasionally brought to San Francisco on vessels in the Gulf trade. It is not common.

Figure 70 in Sowby's Conch. Illustr. without habitat, and named "*C. textile* var." resembles this species. Specimens are in my collection and in that of Mr. Fisher of San Francisco.

This shell belongs to the group of so-called "embroidered cones" of which *C. textile* is the most common illustration, and it might carelessly be mistaken for that species; in *C. textile* however the white (in cleaned specimens) is the dominant color, and the triangular blotches of white are large and sharply defined by a line of brown, and there is but little blending or coalescing of the brown and yellow lines, which are much sharper and more distinct as well as of a lighter shade and narrower than in *C. Dalli*. *C. textile* is of a clear whiteness interiorly, while the shell described herein has a delicate pinkish interior; in *textile* the spire is somewhat *concave*, in *Dalli* it is moderately *convex*; and the latter in outline is a less graceful shell, and belongs to a widely separated zoölogical province.

PTYCHATRACTUS OCCIDENTALIS, Stearns.

P. occidentalis, Stearns, Prel. Descr. August 28, 1871.

Shell elongated, fusiform, rather slender, whitish, traversed by narrow, revolving, brownish threads and much wider intervening spaces; suture distinct, spire tapering; aperture oblong-oval, about half the length of the shell; within white, polished; canal short, nearly straight; columellar obliquely, not strongly plicated; length about three-fourths of an inch.

Habitat.—Near the Island of Nagai, one of the Shumagin Islands, where it was hooked up attached to a rock from a depth of forty fathoms, by Captain Prime of the California Fishing fleet; through the kindness of Mr. Harford to whom it was given, it is now in my cabinet.

This shell in its general features resembles the North Atlantic *P. ligatus* of Mighel and Adams, *vide* Boston Jour. Natl. Hist., IV, 1842. p. 51, pl. iv., fig. 17. It is a more delicate shell than the Atlantic species, though my solitary specimen, judging by the thinness of the outer lip, is not quite mature. I regret that I am unable at present to furnish figures of this and the succeeding species, the specimens having inadvertently been mislaid.

FUSUS (CHRYSODOMUS?) HARFORDII, Stearns.

F. (C.) Harfordii, Stearns, Prel. Descr. August 28, 1871.

Shell solid, elongate, regularly fusiform; spire elevated, whorls six or seven, moderately convex, slightly flattened (in outline) above, with a groove or channel following the suture; color, chocolate brown; surface marked by numerous narrow revolving costæ, which alternate in prominence on the body whorl, and longitudinally by fine incremental striæ, and on the upper whorls by obtusely rounded ribs of more or less prominence; aperture ovate, about one-half the length of the shell, polished, white and finely ribbed within; (the outer lip in perfect specimens is probably finely crenulated); canal short, nearly straight. Lon. 2.1; lat. .94 in. Number of specimens, three; two mature, dead, one junior, fresh.

Habitat.—Coast of Mendocino County, near Big Spanish Flat, California where it was detected by Mr. Harford.

Though almost typically fusiform, except in the brevity of the canal. I am disposed to place it in *Chrysodomus* rather than with *Fusus*. Dr. Carpenter is inclined to believe that certain specimens collected at Monterey by the late Dr. C. A. Canfield and at Catalina Island by Dr. Cooper, are identical with the above. I am of the opinion that it is rather a northern form, exceedingly local in its distribution and more nearly allied to some of the later fossils of the coast described by Mr. Gabb.

PLEUROTOMA (DRILLIA) MONTEREYENSIS, Stearns. Plate I. fig. 2.

P. (D.) Montereyensis, Stearns. Prel. Descr. August 28, 1871.

Shell small, rather solid, elongate, slender; spire elevated, sub-acute; whorls, seven to eight moderately rounded; upper portion of larger volutions somewhat concavely angulated; suture distinct; color, dark purplish brown or black; surface covered with rather coarse, inconspicuous, revolving costae, interrupted on the body whorl by rude incremental lines; middle of upper whorls and upper part of body whorl displaying fourteen to fifteen equidistant, longitudinal, nodose, slightly oblique ribs, which are whitish in the specimen before me (being somewhat rubbed) on the larger whorls; on the smaller volutions of the spire a puckering at and following the suture suggests a second indistinct series of nodules; aperture less than half the length of the shell; canal short; terminal portion of columella whitish, slightly twisted; posterior sinus, rather broad rounded, and of moderate depth. Long. .67 in.; lat. .24 in.

Habitat.—Monterey, California, where the single specimen in my cabinet was collected by Mr. Harford and myself in March, 1868. The shell, in its general aspect, resembles the sombre colored species of the Gulf of California and Panama.

In the cabinet of the Rev. J. Rowell is a specimen perhaps of this species, but not in sufficiently perfect condition to admit of certainty.

PLEUROTOMA (DRILLIA) HEMPHILLII, Stearns. Plate I. fig. 3.

P. (D.) Hemphillii, Stearns. Prel. Descr. August 28, 1871.

Shell small, smooth, slender, polished; spire long, subacute, rounded at apex; longitudinally marked with inconspicuous, oblique ribs, which are nearly obsolete on the body whorl; number of whorls seven, with well defined sutural line, and just below it a parallel impressed thread-like line; shell of an opaque dingy horn color; incremental lines fine, marked in some specimens with dingy white; mouth obliquely ovate, about one-third the length of the shell; labrum produced, anteriorly somewhat thickened; sinus sutural, deep, callonsed; columella thickened at base; canal very short, somewhat produced and twisted; one specimen shows obscure, revolving, impressed lines below the swell of the body whorl; size quite uniform. Long. .26; lat. .09 inch.

Habitat.—Todos los Santos Bay, Lower California, where several specimens were obtained by Mr. Hemphill, for whom I have named this well marked species.

MURICIDEA SUBANGULATA, Stearns. Plate I, fig. 4.

Shell small, abbreviated fusiform, dingy white and marked spirally by an inconspicuous band formed of three reddish-brown lines more or less interrupted on the basal and the preceding volution; whorls five, angulated above and on the basal whorl rounded below the angle, with a shallow sulcation beneath; surface covered with rounded and irregular costæ, which are inconspicuous or obsolete on the upper whorls; longitudinally marked with from seven to nine irregular rounded ribs, which at the edge of the angle (which is somewhat carinated) are broken into angular or pointed knobs or blunt spines; aperture ovate, angulated above and white within; the outer lip with five or six tubercles internally; canal moderately prolonged, slightly curved and open in the two specimens before me. Dimensions of largest: Long. .59; lat. .41 inch.

Habitat.—San Miguel Island, off the southern coast of California, where the specimens from which this description is made were obtained by Mr. W. G. W. Harford.

ASTYRIS VARIEGATA, Stearns. Plate I, fig. 5.

Shell small, elongated, acutely conic, light rufous-brown or sienna-yellow under a thin brownish or greenish epidermis; with whitish median and sutural bands more or less interrupted; in some specimens these bands are connected by waved lines of a darker brown; surface of shell when free from epidermis, smooth and shining, marked with delicate incremental lines, and on the lower portion of the body whorl with narrow grooves; apex rounded, whorls seven, convex; suture well defined, aperture ovate, about one-third the length of the shell; outer lip simple, in some specimens a little thickened with small tubercles on the inner side.

Dimensions: Long. .3; lat. .12 inch.

Habitat.—San Diego, California, where numerous specimens were collected by Henry Hemphill, Esq. This beautiful species resembles some forms of *Nitidella* and *Truncaria*; it differs from *Astyris tuberosa*, in the greater convexity of the whorls, and especially in being without the angularity or concavity which is displayed in the lower part of the body whorl in the latter species; it is a more delicate and graceful shell than either of the other forms of *Astyris* found on the coast, many of which have been distributed as "*Amycla*" or "*Columbella*" *gausapata, Californiana, carinata,* and var. *Hindsii*.

PHOLAS PACIFICA, Stearns. Plate I, figs. 6, 6a, 6b, 6c.

P. *Pacifica,* Stearns, Prel. Descr. August 28, 1871.

Shell oblong, beaks two-fifths of length of shell from anterior end; anterior end of valves triangular, pointed; anterior dorsal edge of valves reflected and folded down on the umbos; lower anterior margin curved, forming a large elliptic-oval gape; posterior end of valves squarely rounded; shell dull chalky white, sculptured in concentric lines, which anteriorly are laminated and posteriorly become extinct; valves radiately ribbed, which also become obsolete at the posterior end; at the intersection of the radiating and concentric lines the sculpture is pectinated; an area below the umbos nearly or quite destitute of sculpture, which varies much in prominence in different specimens; accessory plate sublanceolate and bent down on the beaks, anteriorly prolonged, but not wholly covering the ante-umbonal gape; figs. 6a, 6b, show the variation in the shape of the dorsal plate in different specimens; interior of valves white, enamelled; internal rib short, curved and flattened. Largest specimen, two and six-tenths inches in length, and one and five-tenths inches in height.

Habitat.—Alameda, San Francisco Bay, California, where in some places it is common in sandy mud between tide marks. Numerous specimens collected by Messrs. Harford, Hemphill, Drs. Kellogg and W. P. Gibbons.

This shell is the West Coast analogue of the Atlantic *P. truncata*, Say, which it resembles; it is however a much longer shell for its width, and the portion of the valves posterior to the beaks, very much longer than in Say's species. Specimens of this species have been distributed as *Zirphæa crispata*, which also is found upon the coast, though quite distinct from *P. Pacifica*, which latter comes within Mr. Tryon's subgenus *Cyrtopleura*.

According to the Messrs. Adams in the genus *Pholas*, there are *two* dorsal plates; yet they have included in their list of the species under that genus, *P. truncata*, Say, which has only *one*.

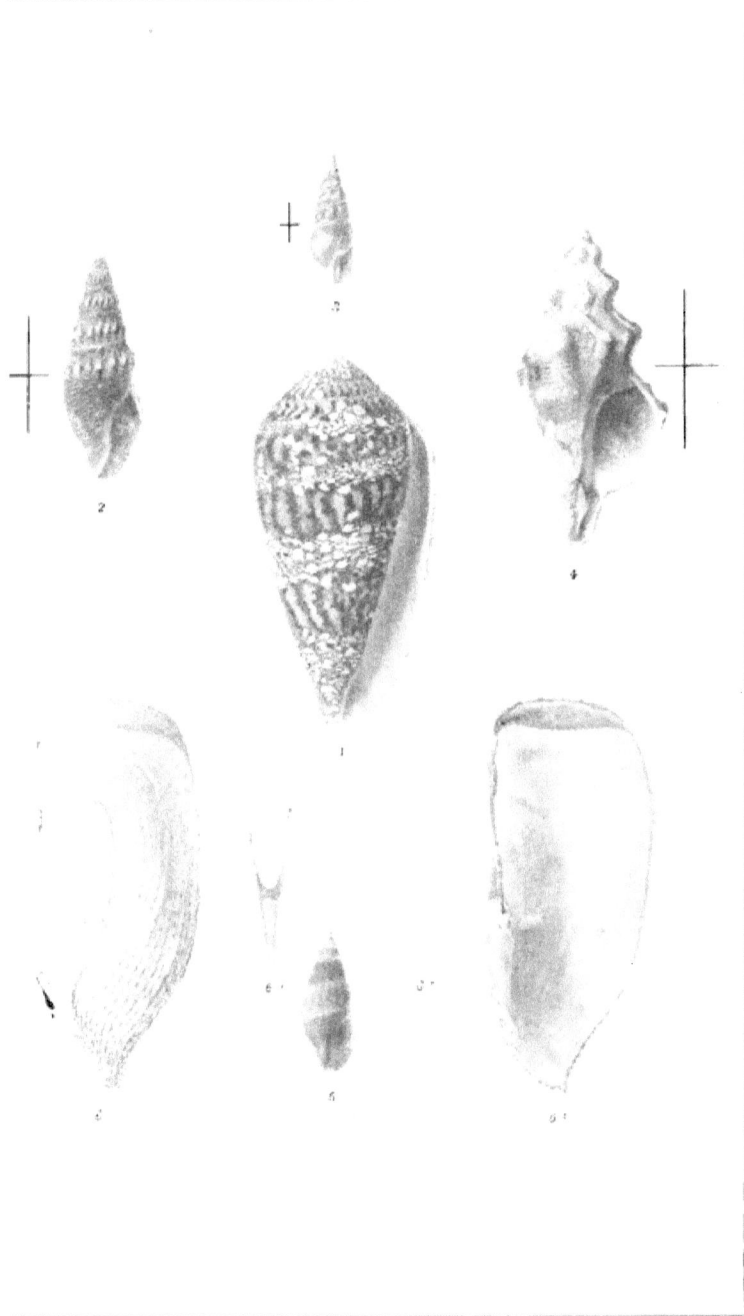

CONCHOLOGICAL MEMORANDA.

No. XIII.

From the Proceedings of the California Academy of Sciences, October 18th, 1875.

On the Vitality of Certain Land Mollusks.

BY ROBT. E. C. STEARNS.

I submit for the inspection of the Academy a living specimen of *Bulimus pallidior*, Sby., one of nine given to me by Prof. Geo. Davidson, who collected them at San José del Cabo, Lower California, in March, 1873.

These snails were kept in a box undisturbed until June 23d, 1875, when I took them out, and, after examination, placed them in a glass jar with some chick-weed and other tender vegetable food, and a small quantity of tepid water, so as to make a warm humid atmosphere. This hospitable treatment induced them to wake up and move about after their long fast and sleep of *two years, two months and sixteen days*. Subsequently all died but this, which seems to be in pretty good health, though not very active.

It may be remembered that I mentioned before the Academy at a meeting in March, 1867, an instance of vitality in a snail (*Helix Veatchii*) from Cerros Island, even more remarkable, the latter having lived without food from 1859, the year when it was collected, to March, 1865, a period of *six years*.

The famous specimen in the British Museum which is cited in the books, *Helix desertorum*, had lived within a few days of four years, fastened to a tablet in one of the cases, when discovered to be alive.

Helix desertorum, as the specific name implies, is found in arid and sterile areas, in the continents of Africa and Asia, and has, as will be perceived, a wide distribution. From the former continent, I have specimens from Egypt, and it also ranges through Arabia in the latter.

The *Bulimus* from the main-land of the peninsula of Lower California, and *Helix Veatchii* from Cerros or Cedros Island, off the coast on the ocean side of the same, come from within the same physical environment, being comparatively a limited distance apart.

The *Helix* belongs to an interesting and peculiar group, probably varieties of one species, which includes, at present, the following names: (1) *Helix areolata*, Sby., (2) *H. Veatchii*, Newc., (3) *H. pandora*, Fbs., and (4) *H. levis*, Ptr. Other forms geographically approximate may hereafter, on further investigation, be referred to the same lineage.

Of the above, (1) *H. areolata* was the first described, or I should say that this appears by the date to be the first name bestowed upon any member of the group. This species has been quoted from Oregon, and (4) *H. levis*, from the Columbia River, in both cases erroneously. The figures in "Land and Fresh Water Shells of North America," * p. 177, are too elevated and globose for the typical *areolata*, but the larger figures faithfully represent *H. Veatchii*. Elevation and rotundity are insular characteristics in this group, and *areolata* is comparatively depressed. It is found in considerable numbers on the uplands around Magdalena Bay, which is on the outer or ocean shore of the peninsula, in latitude about 24° 40' N.

---

* Smithsonian Misc. Coll., No. 194.

Bulimus pallidior, which is pretty generally distributed through Lower California, from Cape St. Lucas northerly, has also erroneously been credited to San Diego in California proper. It is arboreal in its habits, at least during the winter season, and frequents the Copaiva trees. It has been said to inhabit South America, *which is probably incorrect*, and the locality "San Juan," mentioned in "L. and F. W. Shells," on p. 195, where a good figure of this species may be seen, should be *San Juanico*, which is on the *east* side of the peninsula, in latitude about 27° N.

The great importance of particularity in habitat will be at once perceived when I state that there are no less than *three* other localities on the west coast of America, *north* of the place cited, all of which are referred to in various scientific works which have come under my observation, as "San Juan," and there are perhaps as many more "San Juan's" *south* of that especially quoted herein, on the westerly coast of America, in the Central and South American States.

Attention is directed to the fact that the three species herein mentioned as exhibiting extraordinary vitality, belong to geographical areas, which receive only minimum rainfall, or which are, in simple language, nearly rainless regions.

Within such areas vegetation is exceedingly limited even in favorable seasons, and the presence and growth of the annual plants is, of course, dependent upon the rainfall; this last occurring infrequently makes the food supply of land mollusks and other phytophagous or vegetable-eating animals exceedingly precarious.

It is highly probable that a careful investigation in this direction will lead us to the conclusion that the land mollusks which inhabit arid areas have, through selection, adaptation and evolution, become especially fitted for the contingencies of their habitat, and possess a greater degree of vitality or ability to live without food than related forms in what may be considered more favorable regions, and through and by reason of their long sleep or hibernation, *more properly estivation*, with its inactivity and consequent immunity from any waste or exhaustion of vital strength, are enabled to maintain their hold upon life when animals more highly organized would inevitably perish; and we are furnished with an illustration, in the instances cited, how nature works compensatively, when we institute a comparison with the opposite condition of activity, and the food required to sustain it.

I.

II.

1. Bulimus pallidior, *Sby.*
2. Helix Veatchii, *Newc.*

VITALITY OF LAND MOLLUSKS. (STEARNS.)

DESCRIPTIONS OF NEW FOSSIL SHELLS FROM THE TERTIARY OF CALIFORNIA.

BY ROBERT E. C. STEARNS, UNIVERSITY OF CALIFORNIA.

SCALARIA *Lamarck*.

Subgenus **Opalia**, H. & A. Ad.

Opalia varicostata, Stearns. Plate 27. Figs. 2-5.

Shell elongated-conical, turreted, tapering, solid, imperforate, aperture ovate, peristome continuous, thickened; dingy to clear white; suture well defined; whorls united, exceedingly variable in convexity and altitude; specimens all decollate, or truncated, equally solid, though varying in length from .75 to 2.45 inches, showing four and one-half whorls within the first measurement to five in the latter. Perfect specimens have probably from 8 to 12 whorls, or even more. Longitudinal ribs 9 to 12, varying in number, prominence, and regularity, as well as in obliquity, when compared with the axial line of the shell, and, in some specimens, irregularly thickened and distorted by the intrusion of a varical rib more or less conspicuously. In some individuals the termination of the rib at the suture gives the upper part of the whorls a crenulated appearance, and the suture in all specimens is more or less waved, dependent upon the prominence of the ribs, which terminate anteriorly at and join a transverse rib at about the middle of the basal whorl.

Number of specimens 22, all in good condition, save the erosion of the apex.

This is one of those plastic forms which exhibit great variability, but which when a sufficient number of specimens are compared, show well-marked characteristics.

Had the twenty-two specimens examined as above been collected by several persons, and, therefore, divided into many and smaller parcels, and sent, as quite likely would have been the case, to different authors and museums, and thus too widely separated for comparison, it is highly probable, when the latitude of variation which this form presents is considered, that three or four species would have been made out of the above material, which Mr. Henry Hemphill, the collector, kindly placed in my hands for determination.

This is a large species, and perfect specimens, probably, sometimes measure three inches in length; one extreme specimen is strongly suggestive of *Turritella*, and others resemble the living *Opalia borealis*, Gould, common at different places along the coast. It forms a curious, but complete link between the forms like *S. grönlandica*, and the typical *Opaliæ*.

Locality, about eight miles north of San Diego, California, associated with *Pecten* and *Vola*.

Museum of the Smithsonian Institution.

Opalia anomala, Stearns. Plate 27, Fig. 1.

Shell solid, imperforate, elongated-conical, spire gradually tapering; whorls convex, when perfect probably 11 to 14 in number, nearly smooth, being marked only by incremental, and, occasionally, in some specimens, by an outgrown varix; suture well defined; basal whorl traversed transversely by an inconspicuous rib, varying in prominence, in some specimens barely discernible; the convexity or angularity of the lower part of the basal whorl modified by the presence or absence of the basal rib.

Number of specimens 10. Average length 2 inches.

Longitude of smallest . . . . . 1.87 inches
" of largest . . . . . . 2.37 "

As the apex whorls in all of the specimens are wanting, a careful estimate would add .25 inch to the foregoing average, making the latter 2.25 inches in perfect shells.

This species is readily recognized by the absence of longitudinal ribs, though one or two specimens indicate faint plications near the apex.

Locality, the same as the preceding species.

Collected by Mr. Henry Hemphill, of Oakland.

Museum of the Smithsonian Institution.

3

5

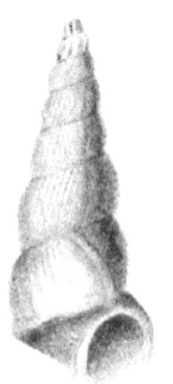

Shells collected at San Juanico, Lower California, by William M. Gabb.

BY ROBERT E. C. STEARNS.

The species contained in this and the succeeding list, were collected in the month of February, 1867, by Prof. Gabb, who kindly submitted the same to me for examination. As the knowledge of Lower California Mollusca is exceedingly limited, the publication of these lists may be of some benefit to students, and of value as data bearing upon geographical distribution. San Juanico is on the east side of the peninsula of Lower California, in latitude about 27° north.

Solecurtus Californianus, *Conr.*
Periploma argentaria, *Conr.*
Standella planulata, *Conr.*
Amphichæna Kindermanni, *Phil.*
Peronæoderma viriditincta, *Cpr.*
Donax flexuosus, *Gld.*
Semele bicolor, *C. B. Ad.*
Fulvia aspersum, *Sby.**
Cyclas dentata, *Wood.*
Mysia orbella, *Gld.*
Crassatella gibbosa, *Sby.*
Lazaria radiata, *Brod.*
Arca Pacifica, *Sby.*
Anomalocardia grandis, *Brod. & Sby.*
    " multicostata, *Sby.*
    " tuberculosa, *Sby.*
Vola dentata, *Sby.*

Chione fluctifraga, *Sby.*
Chione succincta, *Val.*
Chione simillima, *Sby.*
Callista chionæa, *Mke.*
Tivela radiata, *Sby.*
Dosinia ponderosa, *Gray.*
Cardium senticosum, *Sby.*
Omphalius fuscescens, *Phil.*
Crucibulum imbricatum, *Sby.*
Haustator goniostoma, *Val.*
    " tigrina, *Kien.*
Cerithidea albonodosum, *Cpr.*
Trivia radians, *Lam.*
Strombus granulatus, *Swains.*
Neverita Recluziana, *Rve.*
Malea ringens, *Sby.*

*In Adams's Gen. Moll. this species is catalogued as "aspera, *Sby.*;" but in Sowerby's Conch. Illustr. (Fig. 15) it is " C. aspersum, *Sow.*, Zool. Proc. 1833, p. 83," and is credited to "St. Elena, Mr. Cuming." It is strikingly like its Caribbean analogue *F. bullatum*. St. Elena is on the coast of Guayaquil, in latitude about 2 deg. south. If Mr. Cuming's "habitat" is correct, it shows a wide range, and the two species mentioned herein may have descended from the same ancestors.

Chiton (Lepidoradsia) Magdalensis, *Hinds.*
Fissurella volcano, *Rve.*
Callopoma tessellatum, *Kien.*
Lagena nodosum, *Chem.*
Tritonidea insignis, *Rve.*
Cassidulus patula, *Brod.* and *Sby.*

Harpa crenata, *Swains.*
Oliva venulata, *Lam.*
Macron Æthiops, *Rve.**
Fusus Dupetithouarsii, *Kier*
Phyllonotus bicolor, *Val.*
Murex plicatus, *Sby.*

Shells collected at Loreto,† Lower California, by W M. Gab , u February, 1867.

BY ROBERT E. C. STEARNS.

Cyathodonta undulata, *C. r.*
Semele bicolor, *C. B. Ad.*
Chione succincta, *Val.*
Callista chiônæa, *Mke.*
Tapes (Cuneus) grata, *Say.*
Cyclas dentata, *Wood.*
Pecten subnodosus, *Sby.*
Bulla Adamsi, *Mke.*
Acmæa fascicularis, *Mke.*
Crucibulum spinosum, *S g.*
Neritina picta, *Sby.*
Luponia Sowerbyi, *Kien.*
Trivia Solandri, *Gray.*
Surcula funiculata, *Val.*
Architectonica granulata, *Lam.*
Pyrazus incisus (dwarf variety).
Natica Pritchardi, *Fbs.*
Mamma uber, *Val.*

Neverita Recluziana, *Rve.*
Oliva (Ispidula) venulata, *Lam.*
Olivella dama, *Mawe.*
" intorta, *Cpr.*
Purpura (Stramonita) biserialis, *Blainv.*
" triangularis, "
Sistrum carbonarium, *Rve.*
Engina crocostoma, *Rve.*
Columbella fuscata, *Sby.*
Conella cedo-nulli, *Rve.*
Nassa tegula *Rve.*
" versicolor, *C. B. Ad.*
Anachis lyrata, *Sby.*
" nigricans, *Sby.*
" serrata, *Cpr.*
Strombina maculosa, *Sby.*
Murex plicatus, *Sby.*

---

*Macron (a subgenus of *Pseudoliva* made by the Adams's), includes three species all peculiar to the west coast of North America, and inhabiting a semi-tropical and littoral station from (and including) San Diego in California proper, thence southerly, and both coasts of Lower California ; also at " Cedras " or Cerros and other islands along the outer coast of the peninsula ; all of the species are covered with a thick, black epidermis ; *M. Æthiops*, the largest, is traversed spirally by broad, moderately deep grooves from apex to base ; while *M. Kellettii*, A. Ad., has generally only three below the middle of the body whorl, otherwise being nearly smooth. The most northern and smallest of this group is *M. lividus*, A. Ad., which seldom attains the length of an inch, the average of many measurements being .77 inch ; this latter species is proportionately less inflated than either of the others, and is not uncommon at San Diego ; the other species are comparatively rare.

†Loreto is in latitude twenty-five degrees fifty-nine minutes N.; longitude 113 degrees twenty-one minutes W.; Lower California.

face. The teeth above described are on this fish. There are six in three mm.

		M.
Width at shields	. .	.055
" at middle muzzle	. .	.030
Length of head	. .	.045

Prof. Newberry's collection.

PEPLORHINA, Cope.

Established on a species similar to those of the last genus, but with a peculiar sculpture of the scales, which consists of raised points or small tubercles. There is a lateral line of tubes which I cannot find in *Conchiopsis*. An angular bony shield is present behind the gular scutum. There are well ossified ribs, but the structure of the fins cannot be made out at present.

PEPLORHINA ANTHRACINA, Cope.

Scales large, well imbricated; each one is .01 m. in elevation, and three enter .02 longitudinally. The gular and other scuta are smooth, except a band of shallow grooves round the margin.

		M.
Length of gular scutum	. .	.021
" scute behind it	. .	.014

From Prof. Newberry's collection.

DESCRIPTIONS OF NEW MARINE SHELLS FROM THE WEST COAST OF FLORIDA.

BY ROBERT E. C. STEARNS.

ANACHIS SEMIPLICATA, Stearns.

Shell small, solid, elongated-ovate; spire elevated, pointed; whorls 7-8; slightly convex, with inconspicuous revolving grooves, which latter become prominent on the lower portion of the body-whorl; upper portion of the basal, and a portion of the contiguous volution marked by 9-12 moderately sharp longitudinal ribs, which become obsolete on the lower part of the basal whorl, and inconspicuous or extinct on the upper whorls, varying in prominence in different specimens; suture distinct; aperture about half the length of the shell, white, narrow, widest and angulated above; outer lip simple, thickened near the middle, somewhat thickened, shouldered, and curved at its junction with the body-whorl, and nodosely ribbed within; inner lip calloused, callous thin, elevated, and finely tuberculated on its inner edge; canal short, moderately recurved. Number of specimens examined fifteen, of which the largest and smallest measure as follows:—

Fig. 1.

Length .55.   Length of aperture .25.   Breadth .20.
"   .30.   "   "   .15.   "   .12.

Most of the specimens were so coated as not to show the color; when cleansed the surface is of a light sienna-yellow, closely covered with white rounded spots, which frequently coalesce; apex generally eroded.

Habitat.—West Coast of Florida; most abundant at Charlotte Harbor, where Col. E. Jewett collected many specimens, also collected by myself at various points on the shores of Tampa Bay.

The species described herein belongs to a group of Anachids of the same general aspect, of which *A. (Columbella) avara*, Say, is an illustration. *A. semiplicata* differs from Mr. Say's shell in having a greater number of whorls, fewer longitudinal ribs, as well as in color markings, and other minor differences. Through an insufficient comparative examination, specimens have been distributed with Mr. Say's name attached.

ANACHIS ACUTA, Stearns.

Shell small, slender, acutely fusiform; spire elevated, pointed, nucleus rounded, number of whorls eight; in some specimens slightly convex and traversed longitudinally by about fifteen nearly equidistant prominent rounded ribs, which are absent on the apex and adjoining whorl, and become obsolete just below the periphery of the basal volution, which is somewhat angulated below and at its anterior portion marked distinctly with transverse costae; in some specimens the longitudinal ribs show a tendency to nodulation, and terminate rather abruptly upon the periphery of the body-whorl; suture profound, cutting the ribs abruptly; aperture one-third of the length of the shell, whitish, narrowly ovate, angulate above; outer lip simple, moderately thickened; slightly shouldered and curved above, with 5-7 denticles within, which regularly decrease in size anteriorly; inner lip showing a thin polished callus, with, in some specimens, a slightly produced edge; anterior canal short, moderately curved. The shells of this species are quite variable in color, some individuals being of a porcellanous white, with transverse sienna lines and lighter or darker blotches of the same color; others are of a light sienna-yellow, with whitish blotches and brown linear markings. The following are the dimensions of the largest and smallest specimens:—

Fig. 2.

Length .26.   Length of aperture .09.   Breadth .08.
"   .23.   "   "   .08.   "   .09.

Habitat.—West Coast of Florida, at Egmont Key, where several specimens were collected by my friend E. Jewett, Esq. It seems to be rather restricted in its distribution, and much less numerous than *A. avara* or *A. simplicata*; while the number of ribs is the same as in *A. avara*, the shell being much slenderer, these ribs are much nearer to each other, and the surface is destitute of the revolving grooves which characterize that species.

NITIDELLA PILOSA, Stearns.

Shell small, acutely conic, spire elevated, apex rounded; whorls five, convex; suture distinct; surface white, traversed by numerous equidistant fine revolving grooves; body-whorl one-half to three-fifths, and the aperture about one-fourth the length of the shell; mouth ovate, outer lip simple, internally ribbed, thickened at its upper part, turned and reflected upon the body-whorl, forming a callus on the upper portion of the columella, which latter is rather abruptly shortened and slightly twisted anteriorly.

Fig. 3.

Length .16.   Length of aperture .06.   Breadth .06.

Habitat.—West Coast of Florida, on the shores of Tampa Bay, collected by Col. E. Jewett. Only a few specimens were obtained, of which three are in my cabinet. All the specimens examined by me were beach shells, without opercula, consequently the generic position is somewhat doubtful. This species has apparently relations with *Nitidella* (*N. cribraria* or *N. guttata*), also with *Truncaria* (*T. eurytoides*), but as these genera at present include many incongruous species, a revision is necessary before this and similar forms can be satisfactorily placed. The genus *Nitidella* was made by Swainson (vide "Treatise on Malacology," pp. 151–153–313) on the *Columbella nitida*, a West Indian shell, with a short spire, long body-whorl, and a plait on the columella, a markedly different form from *N. guttata* or *N. cribraria*, which are generally included with it. There are numerous allied forms in this connection, which approximate so closely that an examination of the opercula is indispensable to generic determination. (Vide remarks of Mr. Dall in Am. Journ. Conch., vol. vii. p. 115.)

CLATHURELLA JEWETTI, Stearns.

Shell small, abbreviated fusiform, turreted, of a dull ashen-reddish color when dry, dark-reddish to purplish when wet; number of whorls six or seven, convex and somewhat flattened above, sculptured with eleven or twelve prominent longitudinal ribs, which extend nearly or quite to the base of the body-whorl; longitudinal ribs crossed by ten to twelve strong thread-like lateral costæ, which at the points of crossing are in some specimens produced into rounded nodules; the upper whorls show three to four of the lateral costæ, generally three; aperture narrow, about half the length of the shell, of a purplish-black or dark-chocolate color; outer lip thickened and deeply notched above; columella nearly straight, calloused; terminal canal short.

Fig. 4.

Measurements of four specimens as follows:—

Long. .30 inch.   Lat. .14 inch.
"   .28 "   "   .12 "
"   .24 "   "   .11 "
"   .24 "   "   .10 "

Habitat.—Rocky Point, Tampa Bay, West Coast of Florida, where I collected four living specimens from off the underside of clumps of oysters; also at other points along the shores of said bay and the adjacent keys, by Col. E. Jewett, who obtained many specimens. It is perhaps the most common of the pleurotomoid forms of the above region.

August 5.

The President, Dr. Ruschenberger, in the chair.

Ten members present.

August 12.

The President, Dr. Ruschenberger, in the chair.

Six members present.

August 19.

The President, Dr. Ruschenberger, in the chair.

Fourteen members present.

The death of Elias Durand and Dr. L. S. Bolles was announced.

The Composition of Trautwinite. By E. Goldsmith.—The very small quantity of the substance I had for the first examination of the above-named mineral (see Proceedings of the Academy, Jan. 1, 1873) caused me to overlook a few important elements, namely, silica and lime.

John C. Trautwine, to whom my thanks are due for procuring more of the substance, has ascertained that the locality of it is Monterey County, California. The mechanical separation of the Trautwinite from the chromite is a difficult and tedious operation; however, I succeeded so far, that with the lens no black particles of chromite could be discerned.

As the substance is insoluble in acids, I brought it into the soluble condition by fusing it in a mixture of carbonate of soda and saltpetre, treating the fused mass with water until all the soluble parts were exhausted, and the insoluble part with hydrochloric acid. What the acid had not dissolved was again fused with soda and saltpetre, and the obtained mass treated the same as before. From the alkaline solution, after acidulation and reducing the chromic acid to sesquichloride of chromium by hydrochloric acid and alcohol, I separated first silica, then the sesquioxide of chromium.

The other solution, containing the bases as chlorides, was evaporated to dryness, moistened with hydrochloric acid, water added, and thus I found another small quantity of silica. The alumina and iron were then separated from the lime and magnesia by ammonia; but the iron and alumina were again dissolved and

[From the Overland Monthly for April, 1873.]

THE PECTENS, OR SCALLOP-SHELLS.

BY R. E. C. STEARNS.

> The Ocean heaves resistlessly,
> And pours his glittering treasures forth;
> Its waves, the priesthood of the sea,
> Kneel on the shell-gemmed earth,
> And there cast a hollow sound,
> As if they murmur'd praise and prayer;
> On every side 'tis holy ground—
> All nature worships there!
> —Vendor.

OF the many beautiful forms which live in the sea, perhaps none are more attractive or deservedly popular than the pectens, or scallop-shells. The rambler on the sea-shore rejoices in a prize when the odd valve of a scallop is detected in some out-of-the-way nook, covered up and hidden like a treasure, among the sea wrack, mingled in strange confusion, with dead crabs, star-fishes, delicate corals, and algæ—the flotsam and jetsam of the winter storms; and when a specimen of unusual vividness of color and perfectness of sculpture is obtained, an exclamation of triumph mingles with the murmuring music of the surf.

The fairer sex esteem these shells highly, but not from an edible point of view, as do their sterner brethren; for though the animal, or soft part, when *fresh*, is really a great delicacy, the valves, or two pieces of which the complete shell is composed, are utilized in various ways, and with that ingenuity peculiar to the sex, through which "inconsidered trifles" are converted into forms of beauty, an accession of scallops is sure to be followed by a harvest of pincushions and needlebooks.

In natural history, the scallops are known as *Pectens*, from a fancied resemblance of the radiating ribs which most of them display to the teeth of a comb; but as the forms of combs are subject to the caprices of fashion, the pertinency of the name is not altogether apparent. They are also called fan-shells, which is far more appropriate. Though included by the public in the term shell-fish, as are also the clams, quahaugs, and cockles, they are in no way related to the fishes, but belong to the division of the animal kingdom known as mollusca, or soft-bodied animals (from the Latin word *mollis*, soft), as do the cuttles, snails, conchs, oysters, and mussels.

The genus *Pecten* was established by the distinguished naturalist Brugiére, to distinguish these shells from the oysters, with which they were formerly classed. The shells of this genus, of which two hundred species are known, have a wide geographical distribution, being found in almost every sea. In most of them, the valves, as the two pieces are termed which form the perfect shell, are externally convex, but in others one is convex and the other flat. They frequently exhibit most elaborate and exquisite sculpture, and extreme brilliancy of color. One group, which is peculiar to the coral areas of the Indo-Pacific waters, known as the mantle-shells (*Pallium*), resembles fine embroidery in sculpture and coloration. Many of the forms which inhabit the colder seas, either north or south of the equator, are notable for their beauty; a single species frequently indulges in a differentiation in color and markings. The larger species of the fan-shells are found in the colder waters of the North Atlantic and North Pacific (Puget Sound and Japan); also, in the Straits of Magellan, and the similarity of form and sculpture in the shells from these widely separated regions is quite remarkable. Other illustrations of the pectens are found on the west coast of North America, and one species is quite abundant at San Diego.

The fan-shells or scallops were known to the ancients; they were called *Κτένες* by the Greeks, and the *Κτεὶς* of Xenocrates and Galen is said to be the *Pecten maximus* of modern authors. According to Athenæus, this or an allied species was used by the ancients for medicinal purposes as well as food.

In England, they are called "frills," or "queens" in South Devon, according to Montagu; and on the Dorset coast, the fishermen call them "squinns." In the north of France, one kind bears the name of "*vanneau*" or "*olivette*," and another species (*P. maximus*) is an article of food. Of the latter, Jeffrey, a British conchologist, says: "If the oyster is the king of mollusks, this has a just claim to the rank and title of prince." In the fish markets of the north of France, it is called "*grand-pelerine*," "*gofiche*," or "*palourde*." In the south of England, it shares with another species the name of "frill," and in the north that of "clam."

This species (*P. maximus*), Jeffrey says, was formerly "plentiful in Lulworth Bay, on the Dorset coast; but now they are rarely found alive. I was told that the breed had been exterminated there by an epicurean officer of the coast-guard. The late Major Martin would permit any conchologist to dredge as much as he pleased in the bays of the Connemara coast, provided he only took useless shells, ... but all the big clams (*P. maximus*) were reserved for the table at Ballynahinch Castle." The high reputation of this species causes it to be much sought after, and it "is a constant visitant of the London markets. Scalloped with breadcrumbs in its own shell, or fried with a little butter and pepper, it forms a very delicious morsel."

The *Pecten irradians* is the common species on the coast of New England. In the winter the "meats" are sold in Boston market by the quart, and are called "scallops." They are obtained on the shores of Rhode Island. It is somewhat singular that the San Diego scallop has not been introduced into the San Francisco markets; it will be, undoubtedly, in the course of a few years. It may, however, be less palatable than those above referred to, as all the species named inhabit waters that have a much lower temperature during the greater part of the year than the sea at San Diego.

The scallops are, and have been, esteemed for food and other purposes by the aboriginal tribes, as well as by their civilized successors. In the shell-heaps of Florida, among the *Kjœkkenmœddings*, or kitchen-refuse, we find great numbers of these shells, especially in a heap at Cedar Keys; and the shells of some of the west American species, found in Puget Sound, are now used by the Indians of that neighborhood, for in the ethnological department of the Smithsonian Institution at Washington (specimens 4773-4-5) are rattles made of valves of the *Pecten hastatus*, which were used by the Makah Indians in the vicinity of Neeah Bay in their dances;

and another specimen (No. 1034) is a rattle made from the convex valves of a larger species (*Pecten caurinus*) and formerly used as a medicine rattle. These rattles are made by piercing a hole through the valves and stringing them upon a willow, or similar twig.

The animal of the fan-shells is exceedingly beautiful. The mantle, or thin outer edge, which is the part nearest the rim or edge of the valves, conforms to the internal fluted structure of the latter, and presents the appearance of a delicately pointed ruffle or frill. This mantle is a thin and almost transparent membrane, adorned with a delicate fringe of slender, thread-like processes or filaments, and furnished with glands which secrete a coloring matter of the same tint as the shell; the valves increase in size, in harmony with the growth of the soft parts, by the deposition, around and upon the edges, of membranous matter, from the fringed edge of the mantle which secretes it. This cover is also adorned with a row of conspicuous round black eyes (*ocelli*) around its base. The lungs or gills are between the two folds of the mantle, composed of fibres pointing outward, of delicate form, and free at their outer edges, so as to float loosely in the water. The mouth is placed between the two inmost gills, where they unite; it is a simple orifice, destitute of teeth, but with four membranous lips on each side of the aperture.

The pectens have also a foot, less developed than in some others of the bivalve mollusks, which resembles a crooked finger, and is capable of enlargement and contraction, and assists the animal in moving about on the bottom of the sea. Some of them have a sort of beard (*byssus*), at least when young, by which they attach themselves to rocks, seaweeds, and other marine bodies, as do the mussels, which are also bearded; while others of the scallops live without attachment, and move through the water with considerable celerity, with a jerking motion, caused by the rapid opening and closing of the two valves, producing a recoil which carries them along sideways. The young shells of some species dart with great rapidity, a single jerk carrying them several yards. The writer has frequently watched the Atlantic species (*P. irradians*), and when taken from the water, and as long as life continues, the animal will open the valves and shut them with a snap, the operation producing a short, sharp, percussive sound.

The mechanism by which respiration and nutrition are secured is elaborate and exceedingly interesting. The filaments of the gill-fringe, when examined under a powerful microscope, are seen to be covered with numberless minute, hair-like processes, endowed with the power of rapid motion. These are called *cilia*, and, when the animal is alive and *in situ*, with the valves gaping, may be seen in constant vibration in the water, generating, by their mutual action, a system of currents by which the surface of the gills is laved, diverting toward the mouth animalcules and other small nutritious particles.

The shell of the scallops consists almost exclusively, says Dr. W. B. Carpenter, of membranous laminæ coarsely or finely corrugated. It is composed of two very distinct layers, differing in color — and also in texture and destructibility — but having essentially the same structure. Traces of cellularity are sometimes discoverable on the external surface, and one species (*P. nobilis*) has a distinct prismatic cellular layer externally. As the idea of the Corinthian *capital* is believed to have been suggested to Callimachus, the Grecian architect, by a plant of the Acanthus growing around a basket, it is quite possible that the fluting of the Corinthian *column* may have been suggested by the internal grooving of the pecten shells.

Aside from their physiology and the position in the order of Nature occupied by the scallops, they have a place in history and song; for, "in the days when Ossian sang, the flat valves were the plates, the hollow ones the drinking cups, of Fingal and his heroes." The common Mediterranean scallop (*Pecten Jacobæus*), or St. James' shell, was, during the Middle Ages, worn by pilgrims to the Holy Land, and became the badge of several orders of knighthood. "When the monks of the ninth century converted the fisherman of Genneserat into a Spanish warrior, they assigned him the scallop shell for his 'cognizance.'"*

Sir Walter Scott, in his poem, "Marmion," refers to this badge, or emblem, as follows:

"Here is a holy Palmer come,
From Salem first and last from Rome;

---
* Moule's "Heraldry of Fish."

One that hath kissed the blessed tomb,
And visited each holy shrine,
In Araby and Palestine!

* * * * *

In Sinai's wilderness he saw
The Mount where Israel heard the law,
'Mid thunder-dint and flashing leven,
And shadows, mists, and darkness, given.
He shows St. James's cockle-shell—
Of fair Montserrat, too, can tell.
STANZA XXIII.

The summoned Palmer came in place,
His sable cowl o'erhung his face;
In his black mantle was he clad,
With Peter's keys, in cloth of red,
On his broad shoulders wrought;
The scallop-shell his cap did deck."
STANZA XXVII.

And in "The Pilgrimage," written by Sir Walter Raleigh, he says:

"Give me my scallop-shell of quiet,
My staff of faith to walk upon;
My scrip of joy, immortal diet;
My bottle of salvation."

REMARKS

ON

XYLOPHAGOUS MARINE ANIMALS,

OR MARINE ANIMALS WHICH DESTROY WOOD.

BY R. E. C. STEARNS.

At a recent meeting of the California Academy of Sciences, Dr. Hewston submitted specimens of a species of crustacean recently detected in destroying the piling on the water-front of this city, and which it is quite likely is an introduced species, and belongs or is related to *Limnoria* or *Chelura*. A species of the first has for a long time been the source of much expense in various parts of Great Britain, by the damage it causes to the wharf and dock structures of that kingdom.

Mr. Arnold, Civil Engineer, of this city, also exhibited at the same meeting a portion of a pile destroyed by the crustacean referred to, as well as by the ship-worm; the ravages of both, and their method of operating upon the wood, were well illustrated in the specimen submitted.

There are several species of xylophagous or wood-eating marine animals belonging to the Mollusca and Crustacea; included in the former are Teredines (*Teredininæ*) which form a sub-family of the Pholades (Family *Pholadidæ*), and which derive their name from the Greek word *Pholeo*—to bore. The Pholades proper are found in almost every sea, and live in the calcareous rocks and sandstones, as well as in clay and sand, in which they bore their burrows, and specimens are frequently found in some of the harder rock around the Cliff House, near this city. One genus, "*Xylophaga*, is found boring in floating wood, usually forming burrows across the grain about an inch deep, which are oval and lined with shell."*

Of the ship-worms (*Teredininæ*) over twenty are known, and while most of the species in the order in which they are placed, as above mentioned, burrow in harder materials, these confine their labors to wood, boring longitudinally or with the grain, and seldom if ever boring into the burrows of their neighbors. The *Teredines* are divided into three groups. First, *Teredo*; second, *Xylotrya*; third, *Uperotis*. The shell and tube in the sub-genus *Xylotrya* are the same as in the genus *Teredo*; but

---

* Adams' Genera, Vol. II, p. 327.

the shelly processes in *Xylotrya*, or siphonal palettes, as they are called, are long and penniform, or shaped like a quill or feather; to this group belong the three species of ship-worms found in California — *Xylotrya pennatifera*, *Xylotrya fimbriata*, *Xylotrya setacea*. The first and second of the species enumerated are said to be found in England as well as on our coast, while the third species is less widely distributed, and is reported from San Francisco Bay to San Pedro. As before stated, all of the species of ship-worm found thus far in California, belong to the genus *Xylotrya*, and

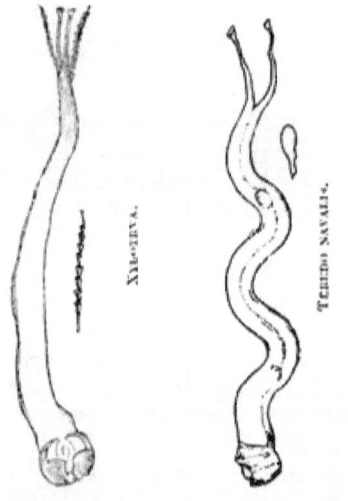

may be distinguished by the two plume-like processes or palettes at the end of the siphons. The destruction caused by the *Xylotrya* (or wood-fish, as the generic word means), is too well known to require comment. I have known a perfectly sound pile of Oregon pine, of a diameter exceeding twelve inches, to be rendered utterly useless in eighteen months by their ravages, while the wood elsewhere than within the section of the pile eaten and bored by them, was as sound and bright as on the day that the pile was driven; this instance is, however, extraordinary. The *Teredo navalis*, to which our ship-worm is frequently referred by local writers, is a foreign species, and the stylets are paddle-shaped, instead of penniform as in the California species.

The third group of the *Teredines* is the *Uperotis*. In these, the shelly valves and palettes are the same, but the tube is twisted and club-shaped from the animal burrowing in the husks of cocoanuts which are found floating in tropical seas.

The ship-worms are ovo-viviparous, the eggs being hatched in the body of the parent, and ejected therefrom through the upper siphonal tube. The young ship-worm, like the young oysters and others of the bivalve mollusks (*Conchifera*), swim freely for a time until they pre-empt or fix upon a pile or other submarine woodwork, when they commence burrowing. While doubtless a great deal of damage is done by them, nevertheless, as is observed by the brothers Adams, "they are useful agents in breaking down and destroying fragments of wrecks and floating timber, which otherwise might be dangerous impediments to navigation."

The ship-worms are frequently two to three feet in length, the body quite soft, but protected by the shelly coating which is deposited by them upon the sides of their burrow, forming a tube, to which, however, in the recent (not fossil) species the body is not attached; the boring is done with the foot, but whether with the shelly valves, which are shaped much like the nibs of a pod-augur, or by some other process, is not by any means a settled question.

In our species the shelly tube is so thin that it is impossible to split the wood without shattering it, but in one species, the *Teredo gigantea*, or giant ship-worm of Linnæus, the tube is often a yard long, and two inches in diameter,"* and exceedingly thick and strong.

The operations of the teredo suggested to that distinguished civil engineer, Mr. Brunel, his method of tunneling the Thames River.†

Having glanced at the xylophagous mollusks, we will now turn briefly to the wood-eating Crustacea, to which the species referred to by Dr. Hewston belongs.

The order of Isopod (or equal-footed) crustaceans inhabit the land and also fresh water and marine stations. As an illustration of their general form we may refer to the terrestrial wood-lice (*Oniscidæ*) which inhabit gardens, cel-

 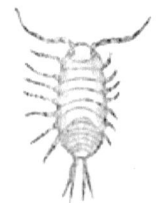

WOOD-LOUSE.   SWIMMING ROCK-LOUSE.

lars, and other damp places, and which are called by the children, "sow-bugs."

The *Isopoda* are divided into three sub-orders: First, the *Ambulatory*, or walkers; second, the *Natatory*, or swimmers; third, the *Sedentary*, or inactive Isopods.

Many of the species are parasites in their habits, and some live in the gills or on the tails of fishes; some live in the bronchial cavity of the land-crab; others in the same portion of the sea-crabs, shrimps, and prawns.

The *Limnoria terebrans*, which is the species so destructive to marine wood-work in Great Britain, belongs to the sea wood-lice, and is the only species in that genus. Whether the form exhibited by Dr. Hewston belongs to this genus or to the genus *Chelura* among

* † Woodward's Manual of Mollusca, 2nd ed., p. 507.

the *Amphipoda*, another order of Crustacea, and which includes a species known as *Chelura terebrans*, remains to be determined. The last-named species was discovered some years ago at Trieste, boring into wood-work in sea-water.

A species of *Chelura* is also found in Australia, and it is not improbable that the species recently found in this harbor may belong to one of the species I have named, and have been imported on the bottom of vessels from some of the ports with which we hold commercial relations.

With the ship-worms and sea-lice operating as in the specimen submitted by Mr. Arnold, a most careful inspection of all submerged wood-work connected with harbor improvements on the coast, is absolutely necessary, or great loss of property and perhaps bodily injury may ensue. So far as protecting piles by saturation with some chemical, obnoxious to the animals referred to in this paper, and which will measurably retain said quality for a reasonable length of time, and resist neutralization by the

LIMNORIA TEREBRANS.   WHITE (SOLDIER) ANT.

sea-water, experiments carefully and considerately made are certainly warranted, and any process by which a successful result should be obtained, would justify a very considerable expense, and still be good financial economy.

Before closing, I will mention another family of wood-eaters, the *Termites*, or white ants, which in some parts of the world are exceedingly destructive; they are principally confined to tropical countries. "When they attack wood-work, they form innumerable galleries, all of which lead to a central point, and in their work they seem carefully to avoid piercing the surface of the wood. Hence the articles which they have perforated appear perfectly sound, when the slightest touch is almost sufficient to cause them to fall to pieces."*

In connection with the depredations of the various forms of *Xylophaga*, it will be well to call attention to certain species of wood referred to in the paper on "Australian Forest Trees," read by me at a meeting of the Academy on the first of last July.† In said paper, I mentioned the *Eucalyptus marginata* (Smith), of which Dr. Mueller says: "The Jarrah or mahogany-tree of S. W. Australia, famed for its indestructible wood, which is attacked neither by *Chelura*, nor *Teredo*, nor *Termites*, and therefore so much sought for jetties and other structures exposed to sea-water, also for underground work, and largely exported for railway sleepers. Vessels built of this timber have been enabled to do without copper-sheathing. It is very strong, of a close grain, and a slightly oily and resinous nature; it works well, takes a fine finish, and is by ship-builders here considered superior to either Oak, Teak, or indeed any other wood." * * * The *E. rostrata* (Schlecht)—the Red Gum of Victoria—is another very valuable species for the "extraordinary endurance of the wood underground, and for this reason highly valued for fence-posts, piles, and railway sleepers; for the latter it will last a dozen years, and if well selected much longer. It is also extensively used by ship-builders. * * * Next to the Jarrah from S. W. Australia, this is the best wood for resisting the attacks of sea-worms and white ants. This species reaches a hundred feet in height.

A consideration of the above facts indicates the course to be pursued. The cultivation of the species of Eucalyptus named herein, would, if commenced immediately, supply us in ten or twelve years with an indestructible timber for submarine wood-work, and prove a profitable enterprise for the capital invested, as well as a great public benefit. But even if a much longer time were required to produce trees of either of these species (whose qualities are so particularly specified by Dr. Mueller) of dimensions suitable for submarine structures, nevertheless the importance of their cultivation is so palpable as to make further comment unnecessary.

---

* Baird's Dict. Nat. Hist., p. 542.
† "On the Economic Value of certain Australian Forest Trees, and their cultivation in California." This valuable paper was printed in full in Vol. II of the CALIFORNIA HORTICULTURIST, pages 271, 313, 326, 355.—ED.

REMARKS

ON THE

Nudibranchiate, or Naked-Gilled, Mollusks.

BY R. E. C. STEARNS.

In the true Gasteropodous (Greek—*gaster*, belly, *podes*, feet), or belly-footed mollusks, we find two great Orders: the Branchifera (from the Latin words *branchia*, gills, and *fero*, to bear), or Gillbearers; and the Pulmonifera (from the Latin *pulmo*, lung, and *fero*, to bear), or Lungbearers—the latter respiring by pulmonary sacs.

Besides this striking difference in structure, the branchiferous Gasteropods pass through a distinct larval stage, and come from the egg in a very different form from that which they present when they are mature, or in the adult stage.

The pulmoniferous mollusks undergo no such metamorphosis.

The *Branchifera* are divided into two sub-orders according to the position of the gills, and the sub-orders are again divided into groups.

One of these sub-orders is called the *Opisthobranchiata* (from the Greek word *opisthos*, behind, and *branchia*, gills), and it is this group to which I refer herein. The second great group of the sub-order *Opisthobranchiata* is the *Nudibranchiata*, or naked-gilled crawlers. The animals, as would be inferred from the name of the group, are destitute of shells, except during the embryonic period, when these delicately constructed creatures are furnished with a small glassy spiral shell, and can swim in the water freely; but as they advance in age the form of the body is modified, and the shell falls off. In the matter of sexual development, they are hermaphrodite; they are also carnivorous in their food, which consists principally of zoöphytes.

The Nudibranchiates are divided into three principal families: the *Eolidæ*, which have the gills along each side of the back; the *Tritonidæ*, which resemble *Eolidæ* somewhat in form and position of the gills, but present other structural differences, which warrant their separation as a family; and the *Doridæ*, generally of broader form and larger size than the two preceding families, as well as of tougher substance, which have the gills placed in a circle on the back and generally in the hinder part of the body. The gills or branchial plumes are of very elegant forms, and frequently present the appearance of Fern-leaves, or similar graceful and feathery shapes; the foot (or belly) is much smaller than the mantle (or back, as seen from above).

The first two families contain species frequently remarkably elaborate in the development of the tentacular processes upon the back and sides, often of most brilliant and varied coloration, while the substance of their bodies is but a trifling degree harder than that of the jelly-fishes; the various tissues of the body being so transparent and delicate that the beating of the heart and the digestive processes are discernible.

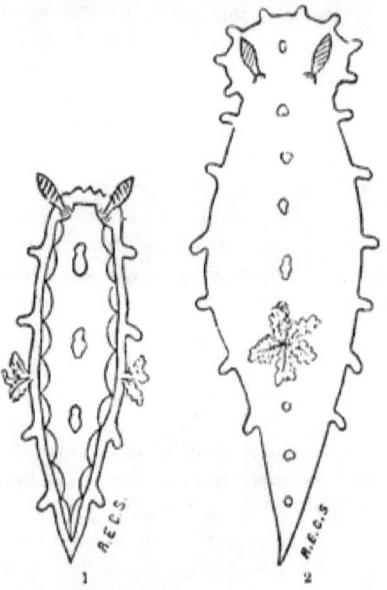

NUDIBRANCHIATE MOLLUSKS.

These remarkable creatures, many of them of marvelous beauty, are found in all parts of the oceanic waters, from the Arctic to the Equatorial seas; probably thousands of species exist as yet undescribed. As but few of the naked-gilled mollusks are of a substance sufficiently solid to admit of preservation in alcohol, they are seldom seen in collections. They may be detected at lowest water-mark on the under side of rocks, appearing to the uneducated observer as nothing more than a highly colored bit of mucus or slime, for the reason that being exceedingly timid, when disturbed they draw their bodies into an almost shapeless lump.

Figures 1 and 2 represent two species of California Nudibranchiate (or naked-gilled) Mollusks, magnified, being twice as long and twice as wide as the living specimens were from which the drawings were made; both belonging to the group of *Triopa*. The first (Fig. 1) is *Laterribranchæa festiva*, so named because the branchæa or gills are on the side of the body and opposite each other; the body is of a transparent cream color, and the festooned or looped lines on the back are of an opaque chalky whiteness, while the substance of the bodies in both of the forms figured is nearly as soft as jelly.

The largest of the two figures is *Triopa Carpenteri*, named for Dr. P. P. Carpenter, a distinguished naturalist, well known among scientific men for his laborious and thorough investigations in the natural history of the west coast of North America. This animal is exceedingly pretty when alive and examined with a magnifier; the upper part of the club-shaped tentacles near the head, and the edges of the gill-plumes which resemble delicate fern-leaves, as well as the ends of the short projecting processes around the edge of the body, are tipped with a brilliant orange, and the body, which is of a translucent whiteness, is covered with fine pimples (*papillæ*) of orange. Both *L. festiva* and *T. Carpenteri* were found on the under side of large granite bowlders near the light-house at Point Pinos, Monterey. When visiting the sea-shore, it will well repay the trouble to turn over some of the bowlders, for Nature hides many such beautiful forms as are above described, in just such out-of-the-way places.

Remarks on Pre-Historic Remains in Florida

BY ROBERT E. C. STEARNS.

Mr. Stearns briefly alluded to his examination and researches, in the year 1869, of the mounds and shell heaps at Point Penallis, Tampa Bay, Florida, near the supposed landing-place of the expedition of De Soto.

These mounds are of two kinds; the mounds proper, being formed of earth, and used for burial and perhaps other purposes, and the others, formed of the shells of mollusks; the latter should, for the sake of distinction and propriety, be called shell-heaps; they are composed of irregularly-alternating, thick strata of shells, and thin strata of ashes; these alternations were owing to a periodicity in the visits of the Indians to the places where these heaps were formed, the selection of a site for a heap being dependent upon the abundance of mollusca of the species considered edible by the Indians, in the adjacent waters; and after the stock of food in the locality was exhausted, it would be left, and the party would remove to another. During the interim between these visits, vegetation, in the form of rank grasses and vines, which are of rapid growth in that country, soon covered the deserted spot; and in the course of several months or a few years, as the case might be, when the locality was again visited, the surface of the heap was burned over, and this being repeated, in the course of years the shell-heaps assumed the form and character in which we find them, and upon being cut through present the appearance of a more or less regular stratification.

The shells found in these heaps are of the same species as are now found living in the neighboring waters; they consist principally of the oysters (*Ostrea Virginica*), conchs (*Busycon perversum* and *Melongena corona*), scallops (*Pecten dislocatus*), and other less common forms.

In the earth or burial mounds, which are of more regular shape and of less size generally than the shell-heaps, were found human remains, which with the exception of the larger bones of the legs or arms, generally crumbled upon exposure to the air, though handled with extreme care. In the shell-heaps and near them, as well as in the burial or earth-mounds, were found fragments of pottery, arrow-heads of chalcedony, and other implements made of stone or shell. Near the Point Pinallis mounds, a remarkable vase made of Steatite, shaped somewhat like a common soup-tureen, had been discovered; through the kindness of Mr. Rothammer, this valuable specimen of aboriginal workmanship was secured by me for the Smithsonian Institution, and is now in its collection; the material of which this vase is made was probably procured at Apalachicola, near which place a deposit of soapstone exists; and the material of which the arrow-heads are made, was, without doubt, obtained at the elevated and now fossilized coral reef in Hillsborough Bay, known as Ballast Point, not far from the town of Tampa.

The great incentive to the invasion and conquest of Florida by DeSoto, was the reputed wealth of the country in gold and pearls; and fabulous accounts were given by the ancient narrators of the abundance of the latter in the possession of the natives or in the hands of the soldiers of that famous expedition. It is highly probable that but few of the rank and file of DeSoto's army were sufficiently familiar with pearls to know anything about them; and if true pearls were found by them in the possession of the Indians, the latter must have obtained them from the freshwater mussels of the rivers, (Unios); and as not one mussel in ten will yield a pearl, and not one pearl in ten will be as large as a pin's head, and not one in one hundred be of proper color, it is much more reasonable to suppose that the so-called pearls were the smooth glossy shells known as *Marginella conoidalis*, which species is quite abundant between tide-marks on the west coast of Florida, and of which a great quantity was found a few years ago in an ancient mound in the city of St. Louis, Missouri. This supposition is strengthened when it is considered how highly certain species of shells are esteemed as ornaments, or for use as money, by barbarous tribes in many parts of the world, even at this day, and the factitious value which is frequently attached to some species of shells by uncivilized man.*

---

*See article entitled "Shell Money," in American Naturalist, Vol. III, pp. 1-5.

Mr. Stearns referred to the early knowledge of the existence of coal oil in the United States, as follows:

The existence of coal-oil in Pennsylvania was known in the last century. In Volume I of the Massachusetts Magazine, published in 1789, by Isaiah Thomas and Ebenezer T. Andrews, names which are historical among the earlier printers and publishers in this country, I find, on page 416, the following:

"In the northern parts of Pennsylvania there is a creek called Oil Creek which empties into the Alleghany River. It issues from a spring, on the top of which floats an oil similar to that called Barbadoes tar; and from which one man may gather several gallons in a day. The troops sent to guard the Western posts halted at this spring, collected some of the oil, and bathed their joints with it. This gave them great relief from the rheumatic complaints with which they were affected. The waters, of which the troops drank freely, operated as a gentle purge."

[From the Proceedings of the California Academy of Sciences, July 1st, 1872.]

ON THE

ECONOMIC VALUE OF CERTAIN AUSTRALIAN FOREST TREES, AND THEIR CULTIVATION IN CALIFORNIA.

BY ROBERT E. C. STEARNS.

Australian forest trees propagated from the seed, with perhaps a few exceptions, thrive remarkably in California; the climate and soil appear to be nearly or quite as favorable to the growth of these exotic as of the native forest forms.

In many of the principal towns in this State, especially in and around San Francisco, in the neighboring city of Oakland and adjoining towns on the easterly side of San Francisco bay, fine specimens of many of the Australian forest species are exceedingly numerous. The most popular of these belonging to the genera *Acacia* and *Eucalyptus*, have been planted for ornamental and shade purposes; the light feathery fern-like foliage of some of the Acacias, their gracefulness, beauty and color combined with rapid growth, present so many advantages as to fairly entitle them to popular esteem. Of the Acacias recommended by Dr. Mueller on account of their economic value,* I am not aware of any being cultivated in this State for that object. *A. decurrens* (= *A. mollissima*) also *A. lophantha* and some other species, are frequent, and highly

* A. decurrens, *Willd*, also A. homalophylla, *Cunn*, and A. melanoxylon, R. Br.

prized for ornamental purposes: from twenty to thirty species are enumerated in the catalogues of the principal nurseries.

The many valuable properties of the species mentioned in the footnote, combined with rapidity of growth, would warrant cultivation on an extensive scale, which if judiciously conducted would be highly advantageous to the State and yield a handsome return upon the capital invested. Mueller says that the wood of A. DECURRENS, popularly known as the "Black Wattle or Silver Wattle," can be used for staves, but its chief use would be to afford the first shelter, in treeless localities, for raising forests. Its bark rich in tannin, and its gum not dissimilar to Gum Arabic, render this tree also important.

A. HOMALOPHYLLA, has a "dark brown wood, is much sought for turner's work on account of its solidity and fragrance; perhaps its most extensive use is in the manufacture of tobacco pipes."

A. MELANOXYLON "is most valuable for furniture, railway carriages, boat building, casks, billiard tables, pianofortes (for soundboards and actions) and numerous other purposes. The fine-grained wood is cut into veneers. It takes a fine polish and is considered equal to the best walnut." Under favorable circumstances it attains "a hight of 80 feet with a stem several feet in diameter." This species requires a deeper and moister soil than *A. decurrens* and *A. lophantha*, which are especially recommended for their ability to resist drought, and therefore particularly applicable to treeless and sterile areas in the southern part of California, and the adjoining country, where the temperature does not decline below 10 degrees.

The peculiar yellow displayed in the China silks and other articles, is obtained from the yellow flowers of a species of Acacia, and is of an exceeding permanent character.

The Acacias are easily propagated from seed, as I have (with some species) practically tested; and it is not unlikely that the flowers of most of the species, which are yellow, might be equally as valuable for the dyer, as the variety cultivated or used by the Chinese.

Of the Eucalypti, E. GLOBULUS is very common in California, and easily cultivated; it is the Blue Gum of Victoria and Tasmania. "This tree is of extremely rapid growth and attains a height of 400 feet, furnishing a first-class wood ; shipbuilders get keels of this timber 120 feet long ; besides this they use it extensively for planking and many other parts of the ship, and it is considered to be generally superior to American Rock Elm. A test of strength has been made between some Blue Gum, English Oak and Indian Teak. The Blue Gum carried 14 lbs. weight more than the Oak, and 17 lbs. 4 ozs. more than Teak, upon the square inch. Blue Gum wood, besides for ship building, is very extensively used by carpenters for all kinds of out-door work, also for fence rails, railway sleepers—lasting about 9 years—for shafts and spokes of drays, and a variety of other purposes." *

Of the rapid growth of this species of Eucalyptus and the facility with which it is propagated, most people in California who have had any experience with it are familiar ; but as perhaps few persons who have specimens of it growing upon their grounds or in their yards are aware of its value otherwise than for ornamental purposes, I have deemed it a matter of interest as well as of importance to quote from Dr. Mueller's valuable paper. Having propagated the Blue Gum from the seed and raised many specimens under not particularly favorable circumstances, I can indorse the remarks of the author from whom I have quoted. An instance of rapid growth immediately under my observation, is that of a specimen purchased by me of a nurseryman, which at the time of planting (Jan. 5, 1871) measured from the ground level to the extreme tip six and one half feet, and in about eleven months (Dec. 8, 1871) had reached a height of a trifle over fifteen feet ; the diameter of the stalk when set out was half an inch, and at the final measurement one and three quarters inches. I am prepared to hear of instances far exceeding my figures, but it should be borne in mind that we had very little rain after this tree was planted, and furthermore that the locality was upon nearly the highest ground in Petaluma. This tree was occasionally, but only moderately watered during a part of the time. Other trees of this species planted at the same time, also made a remarkable growth ; specimens raised by me from the seed, whose growth I have noted, show a gain of ten and a half inches in twenty-one days, or half an inch per diem.

*Vide " The Principal Timber Trees readily eligible for Victorian Industrial Culture, etc., etc., by Ferd Von Mueller."

The development of the lateral branches is as surprising as its perpendicular growth.

George C. Potter, Esq., of Oakland, informs me that specimens upon his grounds nine years old, show a diameter of twelve inches.

Of the large plantation of Eucalyptus of the Blue and Red species made a few years ago by Mr. J. T. Stratton,* of Alameda, I hear indirectly that the trees have done well. I hope at a future meeting to be able to learn from Mr. Stratton, and inform the Academy more definitely of the success thus far, and prospects of this highly commendable and important enterprise.†

The many valuable properties of the Eucalyptus attracted the attention of the French Government several years ago. A specimen in the Jardin d'Acclimation at Algiers, excited the admiration of the Emperor while on a visit to that place, and upon measuring the tree it was found, according to the Paris *Moniteur*, to have made "a height of 30 feet and a diameter of six inches in two years." Since that time it has been extensively cultivated in Algiers, and of late it has been stated that it "is making rapid progress in the south of France, Spain and Corsica, especially on account of its alleged virtues as a remedy for fever. It furnishes a peculiar extractive matter, or alkaloid, called Eucalyptine, said by some to be as excellent a remedy against fever as quinine.

In Spain its efficacy in cases of intermittent and marsh fevers has gained for it the name of "fever tree." It is a powerful tonic and diffusible stimulant, performs remarkable cures in cases of chronic catarrh and dyspepsia, is an excellent antiseptic application for wounds, and tans the skins of dead animals, giving the fragrance of Russia leather. The tree prefers a marshy soil in which it grows to a great height very rapidly. It dries the earth under it by evaporation from its leaves, and shelters it from the sun, thus preventing the generation of marsh miasm."‡

Of the medicinal properties of *E. globulus* we have additional testimony in a recent number of the Practitioner, § where Dr. M. C.

* Report of the Commissioner of Agriculture, 1870, p. 232.
† I do not refer to other forest plantations made in California, by Mr. Aiken or Mr. Edwards, and which I sincerely wish may be successful, for the reason that in this paper the chief object has been to call public attention to certain Australian forms.
‡ Harpers Magazine, March, 1872; Scientific Record, p. 630.
§ No. XLI, p. 368, Nov., 1871.

Maclean relates the results of his experiments on patients in the Hospital Wards at Netley, England. He says in connection with certain cases of chest aneurisms and cardiac asthma, "With the exception, perhaps of the subcutaneous injection of morphia, I know no remedy so efficacious in allaying pain, restoring dyspnœa, calming irritation, and procuring sleep in such cases, as to be compared to *E. globulus*." He also refers to the use in Germany of a tincture made of the leaf, which "has been used successfully in ℨ ij doses in the treatment of intermittent fevers." It appears that it is not only used medicinally in form of a tincture, but also that cigars are made from the leaves, and its palliative influence obtained by smoking.

"German physicians, as appears from medical journals, have found a tincture of the leaves of the *Eucalyptus globulus*, or Australian gum-tree, to be a remedy for intermittent fever. Dr. Lorimer gave it to fifty-three patients, of whom forty-three were completely cured. In five others there was a relapse, owing to a failure in the supply of the tincture. In eleven of the cases quinine had been used without effect, and three of these were cured by the *Eucalyptus*." *

Other species of the Eucalypti, of great value and well worthy of consideration, are recommended by Dr. Mueller.

E. AMYGDALINA, *Labill*, which is sometimes met with 400 feet in height; one specimen in the Dandenong ranges measured 480 feet † surpassing in altitude the gigantic Sequoias of our own State; the wood of this species is said to be well adapted for "shingles, rails, housebuilding, for the kelson and planking of ships, and other purposes;" in rapidity of growth it equals *E. globulus*, but is not so easily satisfied with any soil.

E. DIVERSICOLOR, *F. v. Mueller*, a native of S. W. Australia, sometimes reaching 400 feet in height, with a proportionate growth of stem. The timber is excellent, and young trees are reported as doing well even "in dry exposed localities in Melbourne." It is regarded by Dr. Mueller as a valuable shade tree for avenues, as it makes a dense growth.

The EUCALYPTUS CITRIODORA, *Hooker*, a native of Queensland, "combines with the ordinary qualities of many Eucalypts the ad-

---

* Annual Record of Science and Industry, 1871, p. 586.
† Trans. and Pro. of the Royal Society of Victoria, Part I. Vol. VIII, p. ix.

vantage of yielding from its leaves a rather large supply of volatile oil of excellent lemon-like fragrance."

E. GOMPHOCEPHALA, *Candolle*, grows to a height of "fifty feet, wood close grained, hard and not rending."

EUCALYPTUS MARGINATA, *Smith.* "The Jarrah or mahogany tree of S. W. Australia, famed for its indestructible wood, which is attacked neither by Chelura nor Teredo nor Termites, and therefore so much sought for jetties and other structures exposed to seawater, also for underground work, and largely exported for railway sleepers. Vessels built of this timber have been enabled to do away with copper-plating. It is very strong, of a close grain and a slightly oily and resinous nature; it works well, makes a fine finish, and is by shipbuilders here considered superior to either Oak, Teak, or indeed any other wood." The tree does not grow as rapidly as the Blue Gum in the neighborhood of Melbourne, but Dr. Mueller expresses the opinion that it would make a rapid growth in a more favorable locality.

The E. ROSTRATA, *Schlecht,* the Red Gum of Victoria, is a very valuable species for the " extraordinary endurance of the wood underground, and for this reason highly valued for fence-posts, piles and railway sleepers; for the latter it will last a dozen years, and if well selected much longer. It is also extensively used by shipbuilders, for mainstem, sternpost, innerpost, deadwood, floor timbers, futtocks, transoms, knightheads, hawsepieces, cant, stern, quarter and fashion timber, bottom planks, breasthooks and riders, windlass, bowrails, etc. It should be steamed before it is worked for planking. Next to the Jarrah from W. Australia," this is the best wood for resisting the attacks of seaworms and white ants. This species reaches a hundred feet in height, which is also the height of the next and last of the Eucalypti referred to herein, viz: E. SIDEROXYLON, *Cunn.*, which produces a wood of great strength and hardness, and desirable for carpenters, shipbuilders, and wagon-makers, being suitable for wheels, treenails, belaying pins, and is considered the strongest wood in the colony; also valuable for railway sleepers, underground work in mines, etc.

The wood of the Gums is " so soft at first as to render the felling, splitting, and sawing up of the tree, when green, a very easy process, but when thoroughly dry becoming as hard as oak."*

* Baird's Dict. Nat. Hist., p. 285.

When we consider the fact of the great number of farms in California that are nearly or wholly destitute of wood, and the great and continuous expense entailed by our system of fencing, the importance to the farmer of dedicating a portion of his land to the cultivation of forest trees, from which he can obtain fuel, and fencing materials, is too palpable to admit of debate. The comparatively small expense and labor with which the cultivation of a few acres for the purposes I have named is attended, its absolute feasibility and practicability, with the beneficial results that would flow therefrom, should commend itself at once to every farmer, as a few acres of timber land for economic purposes would add much more than the cost to the cash value of a farm. The boundaries of a farm should be marked by a row or rows of trees, thus defining its limits by living monuments, and greatly adding to its beauty—from these rows as the trees advance in growth and age, some wood could be cut, and where the farm is of considerable size, enough in the way of trimmings or prunings to supply the fuel of the house. In the treeless areas of the southern part of the State, the varieties of Acacia above named would prove an important aid in assisting by their protection the planting of other species of timber; as they are easily taken care of and will stand excessive drouth. They would also be useful as is our Monterey Cypress, (*Cupressus macrocarpa*) for belts to break the force of the winds in exposed places, and it is to be hoped that before many years, timber belts for this purpose will be common wherever the coast winds prevail, as a protection to orchards and vineyards.

We have many native trees well adapted for timber or windbreaks, and while calling the attention of land owners and others to the exotic forms above mentioned and their special qualities as enumerated in Dr. Mueller's excellent paper, I do not wish to be understood as making an unfavorable comparison as against indigenous species, as for some of the purposes mentioned they will answer equally well.

It must be remembered, however, that our forests are unfortunately deficient in many of the hardwoods much used in the arts, and which we are now compelled to import from localities more favored in this respect. The aggregate amount annually sent out of the State for the purchase of this material could by proper foresight and enterprise, in a few years, be retained within our own borders, and here expended in the establishing of new industries pertaining to the very material, the manufacture of which in other portions of the Union employs large communities, to whose support we are now contributing.

As in Germany to anticipate a future need our own *Sequoia sempervirens* or Redwood tree is extensively cultivated, so here by the cultivation of the Australian Eucalypti we can in a few years supply a positive want, and reap the advantages above indicated.

Since the reading of the above paper I have had many questions asked me by persons not present at the meeting of the Academy, and as an answer to said inquiries and to various propositions I have added the following:

Some objection has been made to the Acacias and Eucalypts by persons who have planted them for shade or ornamental purposes in the neighborhood of San Francisco, for the reason as alleged that they do not withstand the winds. So far as the observations of myself and others who have investigated the matter extend, it is really surprising that so few are prostrated. The fault is not with the trees but the purchaser; as trees of from four to six feet in height are sold at a low price, they are bought by parties who require only a few, in preference to smaller trees, as they make a greater immediate show. As most of the growth of the trees as usually purchased, after having attained a height of six inches, has been made in the pot or box in which they are sold by the dealers, it will readily be perceived that the tap-root which in a natural state descends, is diverted from a perpendicular into a rotary direction, analogous to a spiral spring, and is also crossed and recrossed on itself—with the liability as it increases in size to strangle the tree by one portion of this root making a short-turn or twist upon another part of the same, or by being wound about and restricted by the lateral roots. It is therefore apparent that the better policy would be, even where only a few trees are wanted, (and this remark applies with equal pertinence to all trees) that other things being equal, such as comely shape and healthy condition, the younger and smaller trees are really cheaper at the same price than the larger, and can generally be obtained for much less. For forest culture the smaller trees are indispensable to success.

Again it is frequently the case that the lower branches are trimmed off to a mischievous extent, which also is a mistake; for where a tree has sufficient space to grow in, but little trimming is necessary, and it is a false taste which seeks to improve(?) upon nature by depriving a tree of its normal physiognomy and distinctive character by carving it into grotesque or inappropriate shapes; it is simply mutilation, and is certain to result in the premature decay and death of the victim. The flattening of the head by certain aboriginal tribes, and the distorted feet of the fashionable Chinese ladies, are further and pertinent illustrations of analogous hideous violations of natural form.

In compliance with my request to Dr. Arthur B. Stout, of this city, for a relation of his experience with the Eucalyptus in connection with his medical practice, I have received the following:

MR. STEARNS:

Dear Sir: In response to your invitation, I am happy to contribute to your important article on the culture and uses of the Eucalyptus in California, my experience of the medical properties of that valuable plant. The Eucalyptus is not less precious for its medicinal virtues than it is ornamental in arboriculture and useful in the arts. Several months ago, incited by information derived from the *Practitioner* and other sources of knowledge, I collected and dried the leaves. The agreeable empyreumatic oil of the leaves, in evaporating, diffused a balmy odor through the house. I therefore considered that as this oil, as well as the catechu gum and kino, and the cajeput oil, are all similar hydrocarbons, their qualities must resemble the creosote, pyroligneous and carbolic acids in their disinfectant and hygienic properties. I have no doubt that Eucalyptus has these properties in a milder or weaker degree, only differing in being accompanied with an agreeable perfume, wanting to creosote and carbolic acid. As a purifier therefore of the musty atmosphere and unpleasant emanations in basements and cellars, I have recommended the scattering of the dried leaves in such places. The powder of the dried leaves scattered in trunks and among clothes will no doubt be as useful and more agreeable than tobacco or camphor to prevent the growth of moths or other insects. Its chief value is, however, as a sedative and antiseptic in asthma, and throat diseases, nasal catarrhs, and affections of the mucous membranes. To utilize these properties I had a concentrated tincture with alcohol at 95 prepared by Messrs. Steele & Co., and also contrived an inhaler with which to introduce the vapor of the essential oil to the throat and lungs. I can testify to the excellent effect of this mode of medication. The paroxysms of chronic asthma are relieved and shortened, and acute attacks are quickly allayed. The inhaler is a simple instrument made of tin. It is a cup of a capacity of 4 fluid ounces; the lid, attached by a hinge, has a tube from the centre about three inches high, bent near the end at a right angle, and terminated with a mouth piece like that of a speaking trumpet. The cup is on legs so that a spirit lamp may be placed underneath, and has a wooden handle to move it about when heated. Put two ounces of boiling water, (4 tablespoonfuls) in the cup; add one tablespoonful of the tincture; and inhale the vapor, while the fluid is kept gently boiling with the spirit lamp. Again, I had prepared cigarettes with the coarsely powdered leaves. These produce a decidedly anodyne and antispasmodic effect. An agreeable syrup may also be prepared, useful in infantile maladies.

There can be little doubt but that the oil of Eucalyptus, and Eucalyptine when it can be procured, will be available remedies against malarious diseases of all types, and that the presence of the trees, cultivated in gardens, contribute to sanify the atmosphere from those emanations which give origin to epidemic diseases. That the parasitic insects which infest other plants do not relish the Eucalyptus is evident from the general cleanness of the leaves and the fact that the hydrocarbon oils are fatal to animal life. The balmy perfume, therefore, that exhales from them must have an influence in destroying the parasites which frequent shrubs growing in their vicinity, tending to diminish if not suppress them.

In corroboration of the advantages to be obtained by the cultivation of this Myrtacea, may be shown the efforts made during the last fifteen years to acclimate it in Europe and elsewhere. Ramel has succeeded admirably in introducing this tree in Provence (France), in Spain, Italy, the islands of the Mediterranean sea, and in Algeria. It appears in the botanical gardens of Germany (Munich); and in Vienna, Austria, an apothecary, Lamalsh, has raised 3,000 specimens from seeds. From these he has prepared tinctures and oils for medical purposes.

See annual report of Wiggen and Husemann of progress in Pharmacy, etc., Göttingen, 1871.

By the assiduity of Dr. Pigne-Dupuytren, this tree has been carefully cultivated in the garden of the French Hospital of the Mutual Benevolent Association. So, that institution enjoys already the benefit of the tree hygienically, and has its supply of leaves for tinctures and syrups. The leaves steeped in boiling water are also used as a ptisane or beverage.

However obnoxious to parasites in general this tree may be, it appears it nevertheless has its own species in the Psylla Eucalypti. This insect is an Hemipteron, and appears on the Eu. dumosa. It deposits a species of manna, called in Australia *Lerp* or *Laap*. It is a white substance, 53.1 per ct. of sugar syrup and 46.9 p.c. of a special modification of starch. This is prized by the inhabitants as a Manna ; and is greatly sought for by the bees, who convert it into honey. Dobson (entomology) describes it as the cup-like coverings of the Psyllidæ, but Wittstein mentions six varieties of Psylla, and that one species produces a colored Lerp handsomer than the white, but as a deposit beneath the cup like shields of the insect.

See same annual, Göttingen, 1870.

If this insect derives his Lerp from the aromatic and balmy oil of the Eucalyptus, and furnishes an agreeable aliment for the inhabitants, and a Mt. Hymettus-like honey stuff for the bees, certainly the busy little insect manufacturer, parasite as he is, may be freely pardoned. Very respectfully yours,

A. B. STOUT, M. D.

From experiments recently made upon myself, I find that small doses, ℥ ij to ℥ iij, of the infusion of the leaves (of young trees) drank when cold, quiet the nerves and induce sleep ; quite likely, in ordinary cases of wakefulness, a pillow stuffed with the leaves would produce the same result. My friend, Dr. Kellogg, has prescribed the infusion in dyspepsia, and reports favorably. In addition to the many valuable properties of the Blue Gum herein recited, I have no doubt but camphor in considerable quantity can be obtained from it.

A WELCOME TO AGASSIZ.

SAN FRANCISCO, SEPTEMBER 1872

A WELCOME TO AGASSIZ

ALONG six hundred miles of shore,
I heard one day the breakers roar;
A cheery sailor like refrain
It sounded, far across the main,
And like the mists it seemed to rise,
Or tears of joy in friendly eyes,
Until, in one grand peroration,
It burst in hearty salutation!

* * * *

How the animal creation
Had a welcome jubilation,
And the sea-stars and *Echinus*
Said, "We hope that he will find us;"
And the whelks and periwinkles
Laughed their faces full of wrinkles;
And the crabs and lobsters, too,
Drummed on their shells a brisk tattoo,
And the *Octopus punctatus*
Joined the jolly jubilators,
While the festive trap-door spider
Flung his gate a little wider.
And the tree-toads in the willows
And the fishes in the billows,
And the birds among the branches
Of the trees upon the ranches,
All in unison were humming,
A welcome for the Master coming!

The "graylacks" off the Farrallones
Rushed round as if they'd break their bones;
The blackfish and the sleek porpoises,
Were making most exultant noises;
Out at Seal Rocks, the lions there,
Danced a fandango in the air,
Their flippers flapped and wagged their tails,
When they beheld the good ship's sails;
The grizzlies in the hills of Napa
Stood on their heads they felt so happy;
And many a fossil "elefunt"
Felt sad because he couldn't grunt;
The giant clams, great *Schizotharus*,
Were glad to learn that he was near us;
And when the good ship crossed the bar,
They whispered to themselves, "huzza!"

* * * * * * *

How the inanimate creation
Mingled in the jubilation!
The mountains staring at the sea
Were wondering what the muss could be;
But looking down upon the tide,
They felt an earthquake thrill of pride,
For floating lightly on the bay,
The "Hassler" like a sea-bird lay,
And then Mount Shasta said "God bless her!
She's brought us safe the great Professor!"
And from the mammoth trees *Sequoia*,
A voice came, "We've been waiting for you
Some fifteen hundred years or more—
Why haven't you been here before?
The 'Old Maid' now will fall in love!"
Thus spake the Calaveras grove.
The glaciers, pausing on their way,
Said "We've been waiting many a day
And watching anxiously the passes,
To give a greeting to Agassiz."
"Ha!" spoke a smouldering volcano,
"Professor Agassiz? Bueno!"
While far and wide each steaming geyser
Responded with a shrill Aye! Aye! Sir."

How the people sang "Te Deum,"
When they heard who'd come to see 'em!
How everybody there and then
Felt in their hearts a glad Amen,
While e'en Joe Bowers from the grave
Rose up and gave his hand a wave.
And told the story o'er in full,
About the Calaveras skull;
And China John, in soles of wood,
Was heard to mutter, "belly good."

* * * * * * *

A welcome to the genial face,
The generous heart, the friendly hand,
From me and mine to thee and thine
A welcome to the land.
A welcome to the sunset land,
From all its sons and all its daughters;
A welcome to its vine-clad hills,
Its fertile vales and healing waters;
A welcome from its mountain peaks,
A greeting from its wondrous valleys;
A welcome to its mighty groves
And all their shaded alleys.
The golden gate of friendship swings wide open when you come,
And says to you, and Madam, too,—"Now make yourselves at home."

R. E. C. S.

SAN FRANCISCO, September, 1872.

Remarks on the Upper Tuolumne Cañon.

BY ROBERT E. C. STEARNS.

Recent numbers of the *Overland Monthly* have contained contributions by Mr. John Muir, descriptive of the upper valley of the Tuolumne, and that portion of said valley known as the Hetch-Hetchy.

It is gratifying to know that Mr. Muir has found the valley not difficult of access, though at one time supposed to be so, after a partial effort made from an inaccessible point, by Mr. Clarence King.

In the above publication for August last, Mr. Muir says:

"Sometime in August, in the year 1869, in following the river three or four miles below the Soda Springs, I obtained a partial view of the Great Tuolumne Cañon, before I had heard of its existence. The following winter I read what the State Geologist wrote concerning it."

He here quotes from Prof. Whitney as follows:

"The river enters a cañon which is about twenty miles long, and probably inaccessible through its entire length. * * * It certainly cannot be entered from its head. Mr. King followed this cañon down as far as he could, to where the river precipitated itself down in a grand fall, over a mass of rock so rounded on the edge that it was impossible for him to approach near enough to look over. Where the cañon opens out again twenty miles below, so as to be accessible, a remarkable counterpart to Yosemite is found, called the Hetch-Hetchy Valley. * * * Between this and Soda Springs there is a descent in the river of 4,500 feet; and what grand water-falls and stupendous scenery there may be here it is not easy to say. * * * Adventurous climbers * * * should try to penetrate into this unknown gorge, which perhaps may admit of being entered through some of the side cañons coming in from the north."

Mr. Muir here resumes:

"Since that time I have entered the Great Cañon from the north by three different side cañons, and have passed through from end to end, entering at the Hetch-Hetchy Valley and coming out at the Big Meadows, below the Soda

Springs, without encountering any extraordinary difficulties. I am sure that it may be entered at more than fifty different points along the walls, by mountaineers of ordinary nerve and skill. At the head, it is easily accessible on both sides."

I do not intend to question the motive or the taste of Mr. Muir's remarks, which might be regarded as a commentary on his quotation from the State Geologist, or to explain why Mr. King did not explore the valley *at the time* referred to. It seems to me reasonable to suppose that, upon the line pursued by Mr. King, the valley *was* inaccessible; and it is unreasonable to suppose that, if an experienced mountain-climber like Mr. King had really desired to enter the valley, he would have been deterred from doing so by obstacles of an ordinary character, as no person can with truth deny to him the possession " of ordinary nerve and skill."

This interesting region has been again visited this summer by Mr. Muir and several other persons, and will soon become familiar to an increasing number of tourists, from year to year.

On pages 428–9 of Volume I (Geology), in his "Report of Progress and Synopsis of the Field-work" of the Geological Survey, "from 1860 to 1864," Prof. Whitney, in commenting on the main geological and topographical features of this region, remarks:

"The vicinity of Soda Springs, and indeed the whole region about the head of the Upper Tuolumne, is one of the finest in the State for studying the traces of the ancient glacier system of the Sierra Nevada. The valleys of both the forks * * * * * exhibit abundant evidences of having, at no very remote period, been filled with an immense body of moving ice, which has everywhere rounded and polished the surface of the rocks up to the height of at least a thousand feet above the present level of the river at Soda Springs. This polish extends over a vast area, and is so perfect that the surface is often seen from a distance to glitter with the light reflected from it as from a mirror. Not only have we these evidences of the former existence of glaciers, but all the phenomena of the moraines—lateral, medial, and terminal—are here displayed on the grandest scale."

In a foot-note, on page 429, Prof. Whitney says:

"These glacial markings were *first noticed* by Mr. J. E. Clayton, and the fact of their existence was communicated by him to the California Academy of Natural Sciences, *several years ago.*"

(The italics are mine.)

At a meeting of this Academy, held on the 21st of October, 1867,[*] Prof. Whitney exhibited some photographs and stereographs, taken for the Geological Survey by Mr. W. Harris, in the Upper Tuolumne Valley, near Soda Springs, Mount Dana, Mount Hoffmann, and Mount Lyell. He also presented an account of a remarkable portion of the Tuolumne Valley (Hetch-Hetchy Valley), which forms almost an exact counterpart of the Yosemite, written by Mr. C. F. Hoffmann, the head of a party of the Geological Survey, by which it was explored the previous summer.

---

[*] Vide Proc. Cal. Acad. Sci., vol. III, page 368; see also San Francisco *Evening Bulletin* of October 22d, 1867.

REMARKS OF ROBERT E. C. STEARNS

ON THE

DEATH OF DR. WILLIAM STIMPSON,

BEFORE THE

California Academy of Sciences.

June 17th, 1872.

Mr. President:—

It devolves upon me to impart to the Academy the sad news of the death of my esteemed friend, Dr. WILLIAM STIMPSON, late Director of the Chicago Academy of Sciences, and successor to the late Robert Kennicott as scientific head of that institution. Dr. Stimpson was also a corresponding member of this society, as well as many of the principal scientific associations in this country and Europe; and both at home and abroad he was recognized as a most thorough investigator in his department of study, and everywhere his scientific services are appreciated and his loss will be deplored.

From boyhood it may be said, even to the hour of his death, he was a diligent and enthusiastic worker in the enchanting domain of Natural History; first studying with Agassiz soon after the latter settled at Cambridge; at one time shipping as one of the crew on a fishing smack, that he might the better pursue his investigations upon the easterly margin of this continent, especially along the shores of New England. His researches include the entire coast, from Nova Scotia to Key West.

In 1851 he published the "Shells of New England," with notes on their structure, as well as valuable geographical and bathymetrical data; in 1854 appeared his "Synopsis of the Marine Invertebra of Grand Manan or the region about the Bay of Fundy, New Brunswick;" and numerous contributions were made by him to various scientific journals up to the time of his appointment as zoölogist to the United States Surveying Expedition of the North Pacific and Japan Seas, under Commanders Ringgold and Rodgers; (of the results of which expedition a part only has been published, relating principally to Molluscan and Crustacean forms); subsequently his "Review of the Northern Buccinums," Check-list of the Shells of the East Coast of North America from the "Arctic Seas to Georgia," "On the Structural Characters of the so-called Melanians of North America," and "Researches upon the Hydrobiinæ and allied Forms," besides other contributions of more or less importance which have appeared from time to time.

His MSS. relating to the invertebrates of the North Pacific Exploring Expedition, illustrated by numerous drawings, the labor of years, and ready for publication, were all destroyed by the great fire which devastated Chicago, and consumed the building and collections of the Academy of Sciences of that city. Of this terrible conflagration and its result to him, Dr. Stimpson wrote, "My own books, collections, MSS. and drawings—twenty years' work—all gone!"

His health which for some years had been declining, induced him of late to seek, in winter, the milder climate of Florida; and it was upon the eve of one of these winter expeditions to that State (in January to March, 1869) that I had the pleasure of joining and afterward working with him in the field; and of learning, by personal contact, his worth and scientific ability, and his modest estimate of himself.

I shall never forget the delightful season passed in his company, the pleasant toils of each day, and of the rehearsal of each day's triumphs in the evening as we sat in front of the blazing fire of pitch-pine, which lighted up his face with a glow less genial than the smile which played around his lips; or when some joke more pungent than usual was uttered, the explosion of laughter which followed, and which was joined in by none more heartily than himself.

"No hidden snare was in his speech,
Nor malice in his sunny smile."

The destruction of the Chicago Academy's building and its contents involved the loss not only of its own collections, but also much and very valuable material belonging to other institutions which had been sent to Dr. Stimpson to work up. After the disaster and consequent suspension of labor in connection with said Academy, he at once took the field, though in infirm health, in the endeavor to restore as far as possible by new collections the losses which had occurred.

One result of that awful fire, and the most disastrous of all to science, was the shock which it gave to our friend, whose constitution already enfeebled was but poorly prepared for such a blow. In a letter received by me from a mutual friend in Washington, dated December 20th, 1871, he writes: " Stimpson has paid us a call; we have fitted him out, and he has gone to Florida and the West Indies; his health is very bad, he could scarcely look worse and be alive; it is feared he will never regain his health — his severe affliction and great loss by the Chicago fire has too much changed the once energetic Stimpson for him to ever recover, I fear."

In a letter received by me from him, dated September 13th, 1871, (before the fire) he says: " I have a constant cough and difficulty of breathing, with occasional hemorrhage." In his last letter to me, and probably one of the last he ever penned, he said, writing from Key West, April 19th, 1872: " I have been dredging in Coast Survey steamers here and to the westward all winter, but have not got as much as I expected, on account of continued bad weather. * * * My health is very poor — lungs badly filled up with tubercles, etc., and have frequent hemorrhages — cannot do anything requiring any physical exertion without great distress."

But, fellow members, I will not further extend my remarks in reference to the deceased, or enlarge upon his personal merits or scientific ability and services.

From what I have said you will perceive how bravely he worked in the good cause, even in the hour of adversity; with what determination he again went forth to labor, though burdened with disease.

This solemn event, though foreseen, still found us unprepared; and now that it is passed we learn that anticipation cannot shield us from the pain of separation. As we approach his grave we think not less of his high intellectual attainments and scientific ability, we think more highly of the friendship we enjoyed, of the friend that is gone. Never again, save in memory, shall we feel the pressure of his hand, or hear his cheerful voice.

Farewell, dear friend, brave toiler in the glorious cause; we who knew you and loved you will never forget you; in our hearts your memory will ever be green.

If he who makes two blades of grass to grow where but one grew before is a public benefactor, so was he whose death we lament; for his life was a continued effort to increase the sum of human knowledge.

Mr. President, I offer the following resolutions:

Resolved, That the California Academy of Sciences have learned with the deepest regret of the death of Dr. WILLIAM STIMPSON, and deplore the loss of one whose labors in the service of science entitle him to the grateful remembrance of his fellow men.

Resolved, That we extend our heartfelt sympathy to the family and friends of the deceased.

Resolved, That a copy of these Resolutions be forwarded to the family of the deceased, and to the Chicago Academy of Sciences, of which he was a prominent officer.

[From the "Overland Monthly."]

In Memoriam.—William Stimpson, M. D.

BORN, FEB. 14, 1832.—DIED, MAY 26, 1872.

There seems a sadness in the air,
    A shadow on the eastern sky;
The breezes wafted to the west
    Are burdened with a sigh.

The birds are silent in the trees;
    With grief each wild flower droops its head;
The butterflies have furled their wings—
    For Nature whispers, "He is dead."

I walk upon the sea-girt sands;
    A shell is cast within my reach;
Unto my ear I place its lips,
    And listen to its speech.

From far within its heart of pearl,
    A low and saddened undertone—
A blending as of words and tears—
    Says softly, "He is gone."

Dear Mother Earth, within thy breast
    Press tenderly his wasted form;
Sing soothing hymns, ye summer winds;
    Blow gently, winter's storm.

Affection, from thy deepest well
    Renew, with waters sweet and pure,
The freshness of his memory,
    That it may long endure.

                              R. E. C. S.

SAN FRANCISCO, 1872.

Remarks on the Death of Dr. Ferdinand Stoliczka.

BY ROBERT E. C. STEARNS.

MR. PRESIDENT: I regret that I have to announce to the Academy the death of a corresponding member of high scientific reputation and distinguished ability. Dr. Ferdinand Stoliczka, of Calcutta, palæontologist, connected for many years and up to the time of his death with the Geological Survey of India; also Secretary of the Asiatic Society of Bengal, and corresponding member of many scientific societies in Europe and America, and of the California Academy of Sciences since December 18th, 1865.

Dr. Stoliczka was born in Moravia in May, 1838, and died on the nineteenth day of June, at Shayrock, between the Karakorum Pass and Lah in Ladak, while on his return from an exploration amid the mountainous regions of the interior of Central Asia.

He commenced his scientific labors when quite a young man, having joined, soon after finishing his University course, the Imperial Geological Institute of Austria, where he soon displayed great ability as a palæontologist, and by his investigations among the recent and fossil Bryozoa. He joined the British Indian Geological Survey corps in 1862 and worked hard and well in this service, both in the field and the closet, as the publications of the Survey and his many papers in the proceedings of various scientific societies attest.

Dr. Stoliczka's researches were not restricted to the testimony of the rocks, as shown in the numerous fossils described by him; for besides his geological memoirs, his numerous papers on the Natural History of India, including all divisions of animal life, from the higher mammals to the Actinozoa, display his varied knowledge and breadth of study.

His prepossessing appearance, amiable and excellent character, and high culture, gave him a *personnel* altogether attractive, and he was much beloved and esteemed by all who enjoyed the honor of his acquaintance. He died while in the prime of life, in the midst of his scientific labors, not full of years, but nevertheless full of honors.

REMARKS OF ROBERT E. C. STEARNS

ON THE

DEATH OF COLONEL EZEKIEL JEWETT,

BEFORE THE

California Academy of Sciences,

June 18th, 1877.

REMARKS OF ROBERT E. C. STEARNS

ON THE

DEATH OF COLONEL EZEKIEL JEWETT,

BEFORE THE

California Academy of Sciences,

June 18th, 1877.

MEMBERS OF THE ACADEMY: The duty has fallen on me to formally announce to the Academy the death of two of its corresponding members, well-known and highly esteemed in scientific circles--Colonel Ezekiel Jewett and Dr. Philip P. Carpenter. This evening I will read the following brief biographical sketch of the first--reserving a notice of the latter for another occasion.

Colonel Ezekiel Jewett, Ph. D., who was elected a corresponding member of the Academy, April 6th, 1868, was born in the town of Rindge, New Hampshire, October 16th, 1791. His educational opportunities were such as the common schools of the neighborhood afforded at the time. His father, who was a doctor of medicine, would have educated the son for the same profession, but the diverse tastes and restless temperament of the latter, required a broader and more active field of exertion than that of a country physician.

When, in 1812, accumulated grievances culminated in a declaration of war by the United States with Great Britain,

Jewett, then in the vigor of youth, enlisted in the army, and continued in active service until peace was proclaimed. During this military service he was under General Scott, being in the brigade of that celebrated soldier, and received promotion for his gallantry as exhibited on various occasions. He was in the battles of Lundy's Lane, Chippewa and Fort Erie, and at the latter his courage was notably conspicuous. He served his country with distinguished fidelity and bravery, and the commendation of his commanding general was bestowed upon him.

The war with Great Britain being over, and about this time the South American republic of Chili, then a province of Spain, having revolted against the Spanish rule, Jewett, and a few others of his companions-in-arms chivalrously espoused the cause of the Chilians, and pledged their services to the Chilian leader, General Carrera, in behalf of Chilian independence. Crossing the South American continent from Buenos Ayres to Chili, the passage of the Andes was made in a most inclement season and at great peril. When near the crest of the cordillera a fearful snow storm of four days' duration was encountered, in which they nearly perished ; and at the summit, thirteen thousand feet above the sea, the cold was so intense that it was with great difficulty he saved himself from freezing. Arriving in Chili, he took command of the cavalry, and served with distinction until the successful close of the war, when he sailed for Rio Janeiro, and returned to his native village in 1818.

Soon after he married a woman of superior culture and character, Elizabeth Arnold, of Westmoreland, New Hampshire, who proved not only a devoted and affectionate wife, but a sympathetic companion and an appreciative associate in his scientific labors, for whom he ever manifested a most tender regard. In 1826 he removed to Fort Niagara, where he remained in charge seventeen years, his leisure hours occupied with the study of the natural sciences; he also improved the advantages which the locality furnished by making a collection of ethnological material pertaining to the American aborigines. In 1843 he removed to Lockport, New York, and his entire time was now given to the study of geology. In this connection he travelled extensively throughout the United States, including several journeys to the Lake Superior country in the years, 1844, 5, 6, where he was engaged in the exploration of the mineral region, since so famous for its production of copper and iron; and also added largely to his ethnological collection, which he subsequently gave to the Smithsonian Institution. The vicinity of Lockport, at the time referred to, was equal to any, if not the best single point in the world for the student of palæontology. The heavy cutting, through the Niagara limestone, for the locks of the Erie Canal, then in process of enlargement, revealed many wonderful palæontological secrets. It was while Jewett was reaping in this interesting field that he was visited by Agassiz and Ed. de Verneuil, of France, and the acquaintance formed at that time with these eminent men ripened into a friendship which was terminated only by death. At the suggestion of Agassiz, he organized a summer school in geology, which was continued four years, and received the patronage of many now eminent in scientific pursuits. Though Colonel Jewett was especially interested in geology and palæontology, and of material related to the latter had made a large and valuable collection, he was also an eager student in conchology, as will be seen by the following from the report of the British Association for 1863, written by Dr. Philip Carpenter:

"Colonel Jewett went to Panama . . . in January, 1849, spending ten weeks in that region, including Taboga. This was two years before Professor Adams' explorations. Thence he sailed to San Francisco, where he spent four months in exploring the shore for about fifty miles from the head of [entrance to] the bay. After laboring for a week at Monterey, he spent ten weeks at Santa Barbara and the neighborhood, thoroughly exploring the coast for fifteen miles, as far as San Buenaventura. . . Before his return to the East, he also collected at Mazatlan . . . . and Acapulco."

"There can be no doubt of the accuracy of the Colonel's observations at the time they were made. Unsurpassed in America as a field palæontologist, possessed of accurate discrimination, abundant carefulness, and unwearied diligence and patience, no one was better fitted to collect materials for a scientific survey of the coast."

At Santa Barbara he also made a collection of pliocene fossils, which are referred to in the report from which I have quoted. In 1856 he was appointed Curator of the State Museum of New York, at Albany, his incumbency continuing for several years, "a position which he filled," says a writer, "with great credit to himself and incalculable benefit to science."

The voyage from Panama to San Francisco was made on the old whaleship Niantic, as I have learned from a fellow-passenger, a well known citizen of San Francisco. Among the nearly three hundred emigrants, adventurers and pioneers on that now historic vessel—including men of all grades of character and culture, exhibiting every mood and tense of humanity—" Colonel Jewett was a general favorite with them all." . . . " He was a gentleman everywhere and at all times."

In 1859 Numismatics attracted his attention, and with characteristic zeal he pursued the study, and got together in five years one of the largest and most valuable collections of coins and medals in the country.

Early in the beginning of the great civil war, though seventy years of age, he wrote to his old commander, who was then at the head of the army, expressing his readiness to enter the service again, in defence of the Union. The reply of his venerable chief was to the effect, that the magnitude and probable duration of the struggle, required that its burdens and management should devolve upon younger men. During the frequently changing aspect of that prolonged and terrible conflict, those who knew him can tell you how he chafed under this enforced inaction.

PERSONAL REMINISCENCES.

In 1866 he visited California again. It was in the month of June of that year when I met him for the first time and with a few members of the Academy, made up a small party for a short excursion to Bolinas Bay. There are others here to-night who must remember with pleasure the climbing of Tamalpais, the descent to the Bay, and the "walks and talks" with him on that occasion. He was with us but a few months, but sufficiently long to endear himself to all. After returning East, he made several journeys to Florida, during succeeding winters, collecting everything of interest to himself or which might be of service to others. While visiting the East in 1868-70 I again had the pleasure of his companionship on one of these Florida excursions, and with the lamented Stimpson, our little party of three spent the months of January, February and March in the delightful winter climate of that country, collecting along the eastern shore and among the keys on the Gulf side of the peninsula. In November 1869, in company with his friends, Dr. and Mrs. Newcomb, he made a second visit to Panama ; but the climate affected him so severely that he was obliged to return, after a brief stay of only five weeks. He again visited Florida in the winter of 1872, being the fourth time, for the purpose of collecting as before, and was as usual, successful. As may be supposed a man so incessantly active and untiring as Colonel Jewett, was widely known and appreciated in scientific circles, and possessed the friendship and esteem of very many of the most distinguished men of the day.

In 1860 Hamilton College, New York, honored him with the degree of Doctor of Philosophy; and his services to science were further recognized by many learned societies at home and abroad, of which he was an honorary member.

THE CLOSE OF A USEFUL LIFE.

In 1862 he met with the severest affliction in the death of his wife; after this sad event he made his home with his daughter, Mrs. A. A. Boyce, and upon the removal of herself and family to California, about two years ago, accompanied them to Santa Barbara, where, on the 18th of last May, after a brief illness, he closed his eyes forever, at the ripe age of eighty-six years.

Imperfect as is this rapid sketch, it is sufficient to give you some idea of the career of this remarkable man, of his wonderfully active and prolonged life, which exhibited, nearly to its last moments, indomitable energy and perseverance Intellectually of quick perceptions, eager in the pursuit of knowledge, and enthusiastic in his love for and appreciation of nature; actuated by a high sense of honor, and of the most rigid integrity; he was also a man of generous sympathies and impulses. Of exceeding modesty, flattery was distasteful to him, and he was sensitive to the publication of anything in his praise. While courteous to all, he was critical in the selection of his friends, with whom he was exceedingly companionable, and by whom he was greatly beloved.

And here, imperfect as it is, let us close this poor rehearsal of a life well rounded in its fullness of useful service, and of honorable years. Of those refined and delicate qualities which though unseen by the outer world attract kindred spirits and draw them together, I will not speak. Friendship will seek its consolation in precious memories too tender to be told.

REMARKS OF ROBERT E. C. STEARNS

ON THE LATE

DOCTOR PHILIP P. CARPENTER,

BEFORE THE

California Academy of Sciences,

July 2d, 1877.

REMARKS OF ROBERT E. C. STEARNS

ON THE LATE

DOCTOR PHILIP P. CARPENTER,

BEFORE THE

California Academy of Sciences.

July 2d, 1877.

Philip Pearsall Carpenter, Doctor of Philosophy, a corresponding member of this Academy since the year 1863, was born in Bristol, England, on the fourth day of November, 1819, and died at his home, Brandon Lodge, Guy Street, Montreal, C. E., on Thursday morning, May 24th, at seven o'clock.

Doctor Carpenter was one of an illustrious family, distinguished alike for intellectual ability and moral excellence.

His father was the late Rev. Dr. Lant Carpenter, well remembered as an eminent Unitarian divine; his brother, Dr. William B. Carpenter, is one of the foremost scientific thinkers and writers of the day; and his sister, Mary Carpenter, was especially known as an active philanthropist, and of late for her efforts to advance the education of her sex in India. The death of this sister has recently been reported.

Philip Carpenter, as he was familiarly, I might say affectionately called, was educated in Bristol and at the University of Edinburgh; he afterwards studied in a theological institution in the North of England, and combined in a high degree the characteristic virtues of his family.

He entered the ministry, preaching first at Stand near Manchester, and subsequently at Warrington. Ministerial labor, as ordinarily understood was really but a small part of his work. "Universally respected for his ability and general character, he was a leader of the people in everything that was good, and his philanthropy knew no bounds."

He was particularly active in various efforts and ingenious in devising methods for the education and moral elevation of the humbler classes, and established a printing office at Warrington which he called the "Oberlin Press," solely for the purpose of teaching and giving employment to "the youthful poor." From this office, incidental to furnishing instruction and labor, he sent forth pamphlets and handbills denouncing vice and crime; publishing broadcast, sanitary rules and directions for the prevention of pestilential diseases; proclamations of the advantages of ventilation, cleanliness and temperance; culinary instructions, in bread-making and the healthful preparation of simple articles of food; and in other ways with a most righteous zeal, he waged a vigorous and persistent warfare with pen, printing ink and paper, on ignorance and the evils which follow in its train. He was also vehement in his denunciation of slavery; and during the great civil war in America, when many of the cotton mills in England were closed through a lack of the raw material, and the operatives thrown out of work,—unemployed and therefore restless under the pressure of poverty, with a dreary prospect of "no work and no bread,"—he was active alike in devising ways and means by which they could maintain themselves in this emergency—enlightening them by speeches and handbills on the points involved in the contest—thus enlisting their sympathy in behalf of the better cause.

With a taste for Natural History, he commenced the study of the Mollusca, at first as a recreation. "He brought to it a mind well trained in those scientific habits which prevented him from becoming the mere species-monger," and became "one of the most scientific conchologists of our time."

Circumstances placed within his reach a large collection of

shells made at Mazatlan in 1848-50, by Frederick Reigen, a Belgian gentlemen, and believing it conducive to the interests of science, he made a most thorough and critical examination of the abundant material it contained. The result of his labors appeared in the "Catalogue of the Reigen Collection of Mazatlan Mollusca in the British Museum," which he was induced to make by the late Dr. John E. Gray, then at the head of that institution, to which he presented the collection, consisting of eight thousand eight hundred and seventy-three specimens, mounted on two thousand five hundred and twenty-nine glass tablets. He entered upon this work with diffidence, as he remarks in the catalogue, "I undertook the work, trusting that its acknowledged deficiencies might in some measure be compensated for, by great patience and care in the faithful use of those means of information which were within my reach. I have endeavored to make it a companion to Professor C. B. Adams' extremely valuable catalogue of the Shells of Panama, which belong to the same great tropical fauna of Western America."

In the pursuance of his investigations, the variation of species as suggested by intermediate forms, led him to present his views on the importance of study in this direction, and he was requested by the British Association "to prepare a report on the present state of our knowledge with regard to the Mollusca of the West Coast of North America," which was published in the *Transactions* of the Association for 1856, and at once took rank as a most able and conscientious work.

For various reasons—among others his bias for Natural History—he abandoned the ministry, and in 1859, he visited America, "and for some time was engaged in determining and arranging collections of shells presented to the Smithsonian" and other institutions.

The following year he married Miss Minna Meyer, of Hamburg, an estimable lady, who survives him.

In August, 1864, a supplementary report on West American Mollusca was printed, the object of which was as he says, "to correct the errors which have been observed in the first Report," and to present additional information since derived "from fresh sources." This latter was "marked by the same wide range of thought and depth of study" as the first, and has since been re-published by the Smithsonian Institution.

The labor required in the preparation of these two reports was very great, and involved the examination of a vast number of works of travel, records of voyages and expeditions, and the publications of various societies; the examination of numerous museums and private collections and the elaboration of synonomy and the correlation of data scattered here and there in a multitude of volumes, in various public and private libraries.

As one of the many, who have been greatly benefited by Dr. Carpenter's work, I can say with truth, that these conscientiously thorough compilations, made all the more valuable by his judicious comments and methodical arrangement, are of inestimable importance to the student, for they constitute a bibliography of the subject, a starting point and guide for subsequent investigations.

In addition to the foregoing—several monographs of particular groups of shells, and various papers have been published by him from time to time, in the Proceedings of the Zoological Society of London, and of this Academy, as well as in other publications.

In 1865 he removed to America and settled in Montreal. "Here he had hoped to spend his remaining years in his favorite scientific and benevolent pursuits; but shortly after his arrival the failure of a bank in England swept away a large part of the moderate competence on which he relied, and he felt himself necessitated to devote a part of his time to remunerative work. He selected the teaching of boys and persevered in this arduous calling to the end." He was "often urged to abandon his teaching and enter on some line of scientific work more suited to his powers and acquirements; but he preferred his independence and to make his higher scientific and philanthropic occupations altogether labors of love."

His extensive general collection of shells, he presented to the McGill University, and within the last ten years devoted much time to its arrangement, which he left unfinished as well as an elaborate monograph of the perplexing group of the Chitons, which, however, was nearly completed, and will probably be published by the Smithsonian Institution.

Without further ennumerating his manifold scientific labors, and his ever ready assistance to students and others interested in scientific pursuits; his untiring effort in behalf of sanitary and moral reforms, and his attention to other and various duties as a public spirited citizen and neighbor, will cause his name to be held in grateful remembrance in the city of his adoption and by a wide circle of beneficiaries and friends.

A man of broad humanity and of a generous nature, he was full of kindly feeling, and " It is said of him that he could not meet a boy in the street without giving him a loving look." Even once upon a time when his premises were robbed by some unknown person, he published a notice forgiving the thief and inviting him to call and see him.

Scientific men are generally diligent workers and have an extreme appreciation of the value of time. In this respect Dr. Carpenter was noted for his unceasing industry, and his life was without doubt shortened by excessive labor. In a recent letter he kindly inquired for his old friend Jewett, and writes " I am exceedingly over busy."

I will not extend these remarks by a recital of personal reminiscences, or of the many interesting and delightful incidents occurring during several years of familiar intercourse, the remembrance of which will ever be fresh and fragrant as the flowers of spring.

His end was very peaceful, his closing words, a warm farewell. The faith and hope and charity which inspired and guided him through life failed him not in his fading moments. The approach of death was to him but the opening to a glorious day.

REMARKS OF ROB'T E. C. STEARNS,

AND RESOLUTIONS OF THE

California Academy of Sciences,

ON THE DEATH OF

BENJAMIN PARKE AVERY.

California Academy of Sciences.

Regular Meeting, December 6th, 1875.

Mr. President and Members of the Academy:

Since our last meeting the telegraph has brought us sad news—information of the death of our fellow-member, the Hon. Benjamin Parke Avery, United States Minister to China, who died in the early part of November at the city of Peking.

The many excellences of the deceased, the co-operative spirit which he ever manifested in all matters pertaining to the welfare of his fellow-men—quietly, because he was singularly modest and undemonstrative, yet nevertheless persistingly pursuing the even tenor of what he considered his duty—and that duty the advancement of civilization in a new State, the promotion of knowledge, whether in Literature, Science, or Art,—and the general refinement and elevation of the Commonwealth in which he had made his home; such qualities and such services make it eminently proper, that we should inscribe on the permanent records of the Academy, an appreciative recognition of his life and labors, as well as an appropriate expression of our esteem, and of our sorrow for his loss.

With the example of his unassuming but honorable career before us,—too brief but yet well filled with useful work,—it would be in discord with its harmony, to expand these remarks into formal eulogy.

In a letter dated July 5th of this year, the last which I received, he wrote:

"Shut within the walls of our Legation, we are as much alone as if we were in one of the old glacial wombs of the Sierra Nevada—to think of which makes me sigh with longing, for was I not born anew therefrom, a recuperated child of Nature? Your letter with bay-leaves was right welcome, and gave me a good sniff of Berkeley. It was pleasant to receive the University bay, although I am not an Alumnus, and can boast no Alma-Mater except the rough school of self-education."

The closing line above his autograph is "O, California, that's the land for me!" Enclosed with his letter, were a few plants collected by him upon the broad summit of the mouldering walls which surround the ancient city where he died. Our friend has gone—he has found the tranquillity of the grave in a country remote from his native land— from the California he loved so much; far from those he loved and the many who knew and loved him, and who would have deemed it a privilege to have been near him at the final moment, and to have mingled their last farewells with his. The particulars of the closing scene have not yet been received. We may be sure, however, that he looked into the future without fear, and faded serenely, as the twilight sinks into night.

Those who knew him best, and who enjoyed the precious freedom of intimacy will tell you, that his life was conspicuous for its purity—his character for its many virtues—his intellect for its refined and delicate culture—his heart for its tender and generous sympathy. The possession of these qualities endear a man to his fellow men; they constitute a charming whole, whose priceless web is woven from the choicest graces of our poor humanity—they form an enchanted mantle whose shining folds hide the poverty of human limitations.

So lived and walked our friend among us, crowned with the affection and respect of all who knew him. I do not say that he was perfect, and yet if fault he had I know it not, nor never heard it named.

Here let us rest—grateful that so true a life has been a part of ours. We place our tribute on his grave, and say good friend—farewell!

Resolved, That the California Academy of Sciences has learned with profound regret of the death of the Honorable Benjamin Parke Avery, a fellow member and late United States Minister at the Court of Peking ; that we hereby recognize and express our high appreciation of his many private virtues and public services.

Resolved, That these resolutions be spread on the records of the Academy and published in the proceedings.

REMARKS ON A NEW ALCYONOID POLYP,

FROM BURRARD'S INLET.

BY ROBERT E. C. STEARNS.

Remarks on a New Alcyonoid Polyp, from Burrard's Inlet.

BY ROBERT E. C. STEARNS.

At a meeting of the Academy held on the 17th July, 1871 (see Proceedings, Vol. IV, page 180,) in referring to a donation to the Museum, made on the previous 5th of June, of what resembled a bundle "of dried willow switches" from Burrard's Inlet, our fellow member, Dr. Blake, regarded them, as I infer from the brief published abstract of his remarks, as pertaining to a new species of sponge. The exceedingly meagre data in our possession at present, preclude any positive conclusion as to the true position of these apparent "rods or switches of bone," for on referring to our records I see that the specimens were sent "with no information accompanying them, except that they were 'skeletons of some kind of fish!'" At the time of the donation, "It was thought by some to be the internal structure of a species of zoöphyte, allied to Virgularia."

With the specimens alone, and without any knowledge of the fleshy or soft parts, and no particulars as to physiognomy or habit of the organization of which each of these switch-like forms is a part, we can only reason from analogy, and not with satisfactory definiteness.

It is quite certain that they are not the back-bones, and quite unlikely that they are fin-bones of any species of fish; as between zoöphytes and sponges to which latter Dr. Blake regards the specimens as allied, I am decidedly of the opinion, after an examination of the limited authorities at my command, that they belong to a species of zoöphyte, and are included within some one of the groups of the Order of Alcyonoid Polyps.

"The solid secretions of these polyps are of two kinds: Either (1) internal and calcareous; or (2), epidermic, from the base of the polyp. The latter make an axis to the stem or branch, which is either horny * * * o. calcareous. A few species have no solid secretions.

All the species are incapable of locomotion on the base; yet there are some that sometimes occur floating in the open ocean."*

In the third division of the Alcyonoid Polyps, following Prof. Dana's classification, we have the "*Pennatula tribe*, or PENNATULACEA. These are compound alcyonoids, that instead of being attached to rocks, or some firm support, have the base or lower extremity free from polyps and buried in the sand or mud of the sea-bottom, or else live a floating life in the ocean. Their forms are very various."†

After referring to certain species of the *Veretillidæ*, their structure and beauty, other forms are mentioned belonging to the Pennatula tribe, some of

---

* Dana; Coral and Coral Islands, p. 80, 81
† Ibid., page 94.

which, like the group Pennatulidæ, have a stout axis, with branches either side, arranged regularly in plume-like style, or a "very slender stem and very short lateral polyp-bearing pinnules or processes along it (the Virgularidæ); * * * and some of these have a slender stem, and the polyps arranged along one side of it (the Pavonariadæ); and still others a terminal cluster of polyps (the Umbellularidæ).

The most of these species secrete a slender horny axis, and have slender calcareous spicules among the tissues, somewhat like those of Gorgonidæ."*

This internal *horny* axis is also described as "bony"† by other writers: it is covered with a fleshy substance, of a consistence like that of the Actinia, which, being largely composed of water, leaves but little solid matter when dried, which is brushed off or crumbles away with very little handling.

In the Pennatulæ, or Sea-pens, the central stalk or axis is of moderate length and the pinnæ rather long, presenting the appearance of a feather; or as Lamarck said, "it seems, in fact, as if nature, in forming this compound animal, had endeavored to copy the external form of a bird's feather."

"In some genera, *Virgularia* and *Pavonaria*, to which the name of "sea-rushes" has been given, the central stem is very much prolonged, some of them measuring between three and four feet in length. The polypiferous lobes are comparatively short." §

To either the sea-pens (Pennatulidæ), or the Umbellate corals (Umbellularidæ), I believe these specimens belong; and of the two groups indicated, I am inclined to place them in the latter; said group is characterized by a "Polypary free, simple elongated, with the polyps at the summit; axis stony, inarticulate, covered with a fleshy cortex; polyps large, terminal, arranged in an umbellate manner at the end of the polypary."‡

Figuier remarks that "Les Ombellulaires ont une tres-longue tige, soutenue par un os de même longueur et terminee au sommet seulement par un bouquet de polypes."∥

"The physiological phenomena which the Pennatula present is extremely interesting, since it exhibits the example of a truly composite animal, that is, one in which animals, more or less in number, really perfect so far as comports with the grade of organization to which they belong, form part of a common living * * * body, serving as an intermedium for nutrition to all the individuals, so that they are all nourished together in a mediate manner by means of this common portion of which they form a part.

The nutriment which favorable circumstances have placed within the reach of one individual, nourishes that individual first, and then, by extension, nourishes the common stem; and thus the other polypi, which constitute organic portions of it, receive their share."¶

---

* Ibid, page 91.
† Dallas, in "Orr's Circle of the Sciences."
§ Dallas, Ibid.
‡ Manual Nat. Hist. Travellers, page 357.
∥ La Vie et les Mœurs des Animaux, Paris, 1866.
¶ Cuvier; Mollusca and Radiata, by Griffith and Pidgeon. London, 1834.

Or in other words, the nutrition which is secured or received by an individual polyp, is diffused through and nourishes the whole.

After a consideration of the subject, with the specimens before us, I think the analogies strongly favor a reference to one or the other of the groups I have indicated, instead of the fishes or sponges, to either of which I cannot perceive they hold the slightest relationship.

From the coast of Greenland, Lamarck has described a species of Umbellularia (U. Greenlandica.) and we might perhaps, with some degree of reason, look for a related form upon the Pacific side, in some northern station where the physical conditions measurably correspond to those of the habitat of the north Atlantic species cited.

It will be readily perceived, that before an accurate determination can be arrived at, the living forms, of which I believe these "switches" are the central stalks or axes, must be studied *in situ*, as it is quite doubtful whether the fleshy portion can be preserved.

At a meeting of the Academy subsequent to the date of Dr. Blake's remarks to which I have alluded, reference was made to a communication by Mr. Sclater, in the scientific weekly publication, "Nature," bearing upon this subject.

After writing down the conclusions which I have just read, through the courtesy of Dr. Hewston, I was enabled to examine a file of that publication, and I find that Mr. Sclater read a paper before the British Association, at the Brighton meeting, August 20th, 1872,* in which he acknowledges the receipt of several specimens of these "switches," from Captain Herd, of the Hudson's Bay Company, with a statement from the Captain that, "These rods are the backbones of a sort of fish found in great abundance at Burrard's Inlet, Washington Territory, North-west America, whence they have been brought by two Captains in our service. These animals are shaped like a Conger eel, but are quite transparent, their bodies being composed of a mass of jelly — they are about 8 inches in diameter. The head is like a shark's head ; it is attached to the thick end of the rod — it has two eyes and a mouth placed low down. The backbone is also transparent in the living animal, but becomes hard when dried on the beach by the sun. These fishes swim about in shoals, along with the dog-fishes." Other information was received by Mr. Sclater, of the same tenor.

A specimen of the switches was sent by Mr. Sclater to Prof. Kolliker, of Wurzburg, who had shortly before been engaged in monographing the Pennatulidae ; and the latter gentleman, in reply, stated his belief, "That the object you sent me * * * is indeed the axis of an unknown Pennatulidæ, etc."

"Prof. Flower, Prof. Milne-Edwards of Paris, and several other Naturalists, who visited the rooms of the Zoological Society * * all said that the objects were new to them, and that they did not know what they were, but were mostly inclined to regard them as the axis of an unknown Pennatulide animal."†

From the allusion (in the foot-note) in "Nature" to Dr. Gray, and his refer-

---

* See "Nature," Vol. VI, page 136.   † See "Nature"; also foot-note.

ence of one of these switches to a genus (Osteocella) made by him, I quote as follows from page 405, of the Annals and Magazine of Natural History, Vol. IX, (Fourth Series). Dr. Gray refers to the Genus Osteocella as follows: "Mr. Clifton, many years ago, sent * * * to the British Museum, the 'backbone taken out of the marine animal in bottle marked No. 1. I caught him, or it, swimming with great rapidity in shallow water.' The bottle never reached the British Museum; but the backbone did; and I described it at the end of the 'Catalogue of Sea-Pens, or Pennatulidæ, in the British Museum,' published in 1870, under the name of 'Osteocella Cliftoni'; but considered it very doubtful its belonging to the Pennatulidæ."

The British Museum has lately received a very long, slender bone, $64\frac{1}{2}$ inches long and 3-16 inch broad in its broadest part, which was sent to the Zoölogical Society by the Hudson Bay Company, and evidently came from the northern seas, probably from the west coast of America.

Mr. Carter has kindly examined the Australian specimen sent by Mr. Clifton, and the one sent * * by the Hudson Bay Company * * * and finds them, under the microscope, "present the same horny structure, viz., a fibrous trama, more or less charged with oval cells or spaces, quite unlike that of *Gorgonia* and *Pennatula*, which present a concentric mass of horny layers, charged more or less with calcareous crystalline concretions. It is evidently a second species of the same genus, *Osteocella*."

After a few lines, follows a description of the genus

"*Osteocella*, Gray. Cat. of Pennatulidæ (1870), p. 40."

After describing the style, or axis, he refers to the animal (which neither he nor we have seen) in the following words: " Animal or colony of animals free, marine; otherwise unknown; most probably like the Pennatulidæ, but the style is harder, more calcareous and polished than any known style belonging to that group, which are generally square, sometimes cylindrical, but rarely fusiform in the genus *Vergularia*; or, it may be the long conical bone of a form of decapod cephalopod, which has not yet occurred to naturalists, as Mr. Clifton spoke of its being a free marine animal, and it has a cartilaginous apex like the cuttlefish. * * * * It is evident that there are two species of animal yielding this kind of bony substance :

1. *Osteocella Cliftoni*. Thick, about 11 inches long, tapering at each end. From Western Australia.

2. *Osteocella septentrionalis*. Long, slender, about 64 inches long, attenuated at the base, and very much attenuated and elongated at the other end. Northern Seas? Collected by the Hudson's Bay Company."

This latter, undoubtedly refers to the same forms, of which we have numerous specimens in the Academy's Museum, and which are referred to in this paper.

Dr. Gray proceeds and says: "Mr. Carter informs me that subsequent examination of this axis with acid, 'shows that it is similarly composed to that of *Gorgonia*, viz., of kerataceous fibre or substance, and calcareous crystalline matter like that of the stem of *Osteocella Cliftoni*, and the other Pennatulidæ, which it most nearly resembles"; so that my original view as to the nature of this organ seems to be thus confirmed."

From what is herein quoted from Dr. Gray's paper, it will be perceived, that while the microscopic examination showed it to be "quite unlike that of *Gorgonia* and *Pennatula*," that Mr. Carter's subsequent examination of the second species referred to *Osteocella*, "shows that it is similarly composed to that of *Gorgonia*, * * * and * * * like that of the stem of *Osteocella Cliftoni*, and the other Pennatulidae," etc.

Dr. Gray's paper implies a collision between the *microscopic test* and the *examination with acid*; and the description of his genus contains a doubt as to which division of the animal kingdom *Osteocella* is related. With high regard for the justly distinguished naturalist, it must be admitted that his genus is quite indefinite, and could be construed to cover a wide range; but as he has attached it to the catalogue of Pennatulidae, it is perhaps fair to infer that in his mind the balance of reasoning tends in that direction; as between the microscopic and the acid tests, the latter is of insignificant value.

But returning to the " switches," I find that Mr. Sclater does not commit himself, but with apparent consideration for the intelligence of the parties who sent him the specimens and their statement that they belonged to a species of fish, he only says that, "supposing * * * * that these objects are really derived from such an animal as is described and figured above, I can only suggest that they may be the hardened notochords of a low-organized fish, allied either to the Chimaeroids or to the Lampreys, in which the notochord is persistent throughout life. It is quite certain, I think, that they cannot be any part of the true vertebral column."

On page 432 of the same number of "Nature," appears an article relating to Mr. Sclater's paper, from Mr. H. N. Moseley, who, after what appears to have been a rather careful examination of the authorities upon the groups to which he thinks it belongs, as well as upon its microscopical structure, expresses an endorsement of Prof. Kolliker's opinion, and closes by saying : " In the mean time I cannot but conclude that Mr. Sclater has been misinformed, and that we are very unlikely ever to see that marvellous fish in the flesh."

Again : in " Nature," of October 24th, 1872,* Mr. J. W. Dawson, Principal of the McGill College, at Montreal, writes that, presuming that the " disputed organism * * is specifically identical with a specimen from Frazer River * * presented * * for the Museum of the University * * *. I at once recognised it as the axis of a Virgularia, or some similar creature * * * *. I submitted it to Prof. Verrill, of Yale College, who had no doubt as to its nature;" and Mr. Whiteaves, of Montreal, noticed it in his report, " as an undescribed Pennatulid."

Then follows Dr. Blake, in " Nature," (of November 28th, 1872)† to which previous reference has been made by me, as it is a part of this Academy's proceedings, in which, as the result of a microscopic investigation, he says : " An

---

* Vol. VI, No. 156.   † Vol. VII, page 161.

examination of the specimens * * enables me to refer them to the Protozoa class, Spongidæ, or sponges"; and he concludes by saying : " Its generic relations will, I think, be with Hyalonema and Euplectella, both sponges of the Pacific."

The foregoing is all that I find relating to the "switches," prior to my remarks as above; I was not aware, at the time, that anything had appeared on the subject, other than the remarks of Dr. Blake, and that of Mr. Sclater's article, to which Dr. Blake referred. Mr. Sclater's article I had not read, but had casually glanced at the drawing of the so-called fish.

But having expended so much time prior to an examination of the files of "Nature," I considered it a matter of sufficient interest to warrant a review of the subject, and present the same to the Academy.

As to what these animal "switches" belong, it will be seen that Dr. Blake, whose examination of their substance microscopically appears to have been quite thorough, places them with the *sponges*. Mr. Sclater does not commit himself, but *conditionally* refers them to the *fishes*. Dr. Gray described (it) them as a new species of *Osteocella*, whatever that may be, (perhaps a *Pennatulid*) while Professors Kolliker, Flower, Milne-Edwards, Mr. Mosely, Principal Dawson, Prof. Verrill, Mr. Whiteaves, Mr. Dall and myself, regard them as belonging to a species of Alcyonoid polyp, related or pertaining to the group *Pennatulidæ*.

On reviewing the above, it will be noticed that the various parties who presented the specimens, both of the Burrard's Inlet forms and that from West Australia, state that they are bones of, or belong to fishes, implying that they are a part of free-swimming animals; while some species of the *Pennatulacea* "live a floating life in the ocean," it is not unlikely that others may not be constantly stationary, or, if I may use the word, are not *planted, all* of the time ; and while floating might be mistaken for fishes, more especially if numerous specimens were seen moving in the water, coincident with the presence of a school of fishes.

In conclusion, I would state my belief that the much-discussed switches are a species of *Umbellularia*, for which Dr. Gray's specific name might be adopted, and attached to the specimens from Burrard's Inlet, in the Academy's collection.

[From the San Francisco "Mining and Scientific Press," August 9, 1873.]

Description of a New Species of Alcyonoid Polyp.

BY ROBERT E. C. STEARNS.

At a meeting of the California Academy of Sciences, held on the third day of February, 1873, a paper was read by me, entitled "Remarks on a New Alcyonoid Polyp, from Burrard's Inlet;"* in which I gave a *resumé* of the discussions, notices, etc., in this country and in England, arising from the examination by several naturalists, of certain "switch"-like forms, which had been received by different parties from the Gulf of Georgia (more particularly from Burrard's Inlet, in said gulf); several specimens of said "switches" being in the Museum of the California Academy.

These "switches," or rods, were referred by Dr. Gray, of the British Museum, to his genus "Osteocella," and by Mr. Sclater's correspondent stated to belong to " a sort of fish ;" but by the majority of scientific gentlemen who had seen these "switches" they were regarded as belonging to a species of Alcyonoid Polyp. I expressed the belief that they belonged to a species of *Umbellularia*.

At a meeting of the California Academy, held on the evening of August 4, 1873, Dr. James Blake presented a specimen of the polyp of which these so-called switches are the axes, which had been sent to him from the Gulf of Georgia by his friend, Capt. Doane. This specimen was one of six or seven sent at the same time, all of which were in a tolerable state of preservation, though, as might have been anticipated, the more delicate tissues of the polyps are somewhat decomposed, and some of the specimens are in some places lacerated. They all are, however, sufficiently perfect to determine the true position, and show that the "switches" are, as was supposed, the supporting stalks or axes of an Alcyonoid Polyp "related or pertaining to the group *Pennatulida*."

On the day after the meeting of the Academy (August 5), through the courtesy of Dr. Blake, I was invited to inspect the other specimens, and from said examination have written the following description :

Genus PAVONARIA, Cuvier.

Pavonaria Blakei ; n. s.

Polyp-mass or polypidom, of a flesh or pink color, linear, elongate, attenuate ; polypiferous portion about three fourths of the entire length, rounded oval to ovate elliptic in cross section, and from three fourths to one inch in greatest di-

---

* *Vide* Proceed. Cal. Acad. Sciences, Vol. V., Part I., pp. 7-12.

ameter, flatly tapering toward the tip, as well as decreasing in the opposite direction to where the polypiferous rows terminate or become obsolete. From this latter point to the beginning of the base or root, a portion of the polypidom, equal to about one sixth of its entire length, is quite slender, being only about twice the diameter of the naked axis, and the surface quite smooth; said portion, as well as the base, is round (in cross section); the basal part is from one ninth to one eleventh of the entire length, and about one inch in diameter, with the surface longitudinally wrinkled or contracted, presenting a ridged or fibrous appearance.

Style or axis long, slender, white, hard, bony, somewhat polished, about three sixteenths (3-16) of an inch in diameter in the thickest part, tapering gradually toward the tip, and attenuated, with surface somewhat roughened toward the basal extremity. Inclosed in the polyp-mass or polypidom, the axis is central from the base to where the polyp-rows begin, when it soon becomes marginal or lateral, forming a prominent rounded edge (free from polyps) on one side of the polypiferous portion of the whole. From near the sides of the axial edge the polyp-rows start, and run obliquely upward to the opposite side, where they nearly or quite meet, presenting, when that side is observed from above, a concentric chevron or $\wedge$-like arrangement. The more conspicuous polyp-rows show from nine to fourteen polyps, with occasional intermediate rows of three or more polyps.

The length of the most perfect of Dr. Blake's specimens was sixty-six (66) inches; of which, commencing at the tip, a length of forty-eight and a quarter ($48\frac{1}{4}$) inches was occupied by the polyp-rows, which numbered two hundred and forty-five (245), or twice that number when both sides or arms of the chevron or $\wedge$ are considered. The number of polyps in each row was, in this specimen, from eight (8) to eleven (11), with occasional intermediate shorter rows of from three (3) to seven (7). Estimating ten to the row, this specimen exhibited about *five thousand* polyps, all of which, as well as the polyps in the other specimens, were filled with diminutive ova, of an orange color. In the next section of this specimen, the length between the last polyp-row and the swell of the base or root, is eleven and one quarter ($11\frac{1}{4}$) inches; thence to the termination of the base, six (6) inches.

The average dimensions of thirty-six (36) of the axes in the Museum of the California Academy is five feet six and one third inches in length, and the diameter or the largest, nine thirty-seconds of an inch; diameter of smallest specimen, one sixteenth of an inch.

In connection with the above description, some allowance should be made for the contraction and injury of the tissues by the alcohol in which the specimens were placed after they reached this city.

Additional specimens of the above species, from the same locality, have been received from J. S. Lawson, Esq., U. S. Coast Survey, by George Davidson, Esq., President of the Academy.

APPENDA.—Of the specimens received from Mr. Lawson, some individuals are younger than either of Dr. Blake's. In these the polyp-rows are farther apart, and there are not so many polyps in the row; neither do the ends of the rows approximate so closely on the side opposite the axial edge; the polyps being not nearly so many in the same length, or presenting (as do some of Dr. Blake's specimens) so crowded an appearance. In cross-section through the polypiferous portions, the younger individuals are less ovate or acutely oval than in the older specimens. The general aspect of this species, judging from the figure in Plate XXXI. of Johnson's British Zoophytes (2d ed.), is like *P. quadrangularis* from Oban.

Description of a New Genus and Species of Alcyonoid Polyp.

BY ROBERT E. C. STEARNS.

At a meeting of the California Academy of Sciences, held on the third day of February, 1873, a paper was read by me, entitled "Remarks on a New Alcyonoid Polyp, from Burrard's Inlet;"* in which I gave a *resumé* of the discussions, notices, etc., in this country and in England, arising from the examination by several naturalists, of certain "switch"-like forms, which had been received by different parties from the Gulf of Georgia (more particularly from Burrard's Inlet, in said gulf); several specimens of said "switches" being in the Museum of the California Academy.

These "switches," or rods, were referred by Dr. Gray, of the British Museum, to his genus "Osteocella," and by Mr. Sclater's correspondent stated to belong to "a sort of fish;" but by the majority of scientific gentlemen who had seen these "switches" they were regarded as belonging to a species of Alcyonoid Polyp. I expressed the belief that they belonged to a species of *Umbellularia*.

At a meeting of the California Academy, held on the evening of August 4, 1873, Dr. James Blake presented a specimen of the polyp of which these so-called switches are the axes, which had been sent to him from the Gulf of Georgia by his friend, Capt. Doane. This specimen was one of six or seven sent at the same time, all of which were in a tolerable state of preservation, though, as might have been anticipated, the more delicate tissues of the polyps are somewhat decomposed, and some of the specimens are in some places lacerated. They all are, however, sufficiently perfect to determine the true position, and show that the "switches" are, as was supposed, the supporting stalks or axes of an Alcyonoid Polyp "related or pertaining to the group *Pennatulida*."

At the last meeting I referred the specimen before the Academy to that division of the *Pennatulida* known as *Virgularia*, but upon a subsequent examination of the authorities, I find that those forms in which the axis is unilateral, or on one side, come within the Genus *Pavonaria* of Cuvier.

The only species heretofore described so far as I can learn, and on which this genus is based is *P. quadrangularis*, of which a lengthy and interesting description from Prof. Forbes, is given in Johnston's British Zoöphytes (Vol. I, pp. 164–166). In that species however, the axis is "acutely quadrangular" and the polyps are arranged in three longitudinal series, corresponding to three of the "angles of the stem."

---

* *Vide* Proceed. Cal. Acad. Sciences, Vol. V, Part I, pp. 7–12.

In the specimen presented by Dr. Blake the style or axis is round and the polyps are arranged in two longitudinal unilateral series, which conform to the convexity of the external fleshy covering. With these differences, I think I am justified in placing it in a new sub-genus for which I propose the name of *Verrillia* in honor of Prof. Verrill of Yale College.

Genus PAVONARIA, Cuvier.
Sub-genus VERRILLIA, Stearns.

Polypidom linear-elongate, round, oval or ovate in cross-section. Axis round, slender, bony; polyps arranged in two unilateral longitudinal series.

Verrillia Blakei, Stearns; n. s.

Polyp-mass or polypidom, of a flesh or pink color, linear, elongate, attenuate; polypiferous portion about three fourths of the entire length, rounded oval to ovate-elliptic in cross section, and from three fourths to one inch in greatest diameter, flatly tapering toward the tip, as well as decreasing in the opposite direction to where the polypiferous rows terminate or become obsolete. From this latter point to the beginning of the base or root, a portion of the polypidom, equal to about one sixth of its entire length, is quite slender, being only about twice the diameter of the naked axis, and the surface quite smooth; said portion, as well as the base, is round (in cross section); the basal part is from one ninth to one eleventh of the entire length, and about one inch in diameter, with the surface longitudinally wrinkled or contracted, presenting a ridged or fibrous appearance.

Style or axis long, slender, white, hard, bony, somewhat polished, about three sixteenths (3-16) of an inch in diameter in the thickest part, tapering gradually toward the tip, and attenuated, with surface somewhat roughened toward the basal extremity. Inclosed in the polyp-mass or polypidom, the axis is central from the base to where the polyp-rows begin, when it soon becomes marginal or lateral, forming a prominent rounded edge (free from polyps) on one side of the polypiferous portion of the whole.

From near the sides of the axial edge the polyp-rows start, and run obliquely upward to the opposite side, where they nearly meet, presenting, when that side is observed from above, a concentric chevron or $\wedge$-like arrangement, modified by the convexity of the polypidom. The more conspicuous polyp-rows show from nine to fourteen polyps, with occasional intermediate rows of three or more polyps.

The length of the most perfect of Dr. Blake's specimens was sixty-six (66) inches; of which, commencing at the tip, a length of forty-eight and a quarter (48$\frac{1}{4}$) inches was occupied by the polyp-rows, which numbered two hundred and forty-five (245), or twice that number when both sides or arms of the chevron or $\wedge$ are considered. The number of polyps in each row was, in this specimen, from eight (8) to eleven (11), with occasional intermediate shorter rows of from three (3) to seven (7). Estimating ten to the row, this specimen exhibited about *five thousand* polyps, all of which, as well as the polyps in the other specimens, were filled with ova, of an orange color. In the next section of this specimen, the length between the last polyp-row and the swell of the base or root, is eleven and one quarter ($11\frac{1}{4}$) inches; thence to the termination of the base, six (6) inches.

The average dimensions of thirty-six (36) of the axes in the Museum of the California Academy is five feet six and one third inches in length, and the diameter of the largest, nine thirty-seconds of an inch; diameter of smallest specimen, one sixteenth of an inch.

Dr. Blake's specimens were preserved in a mixture of glycerine and alcohol, and the more delicate tissue of the polyps appears to have been somewhat injured by the latter ingredient.

Additional specimens of the above species, from the same locality, have been received from J. S. Lawson, Esq., of the U. S. Coast Survey, by George Davidson, Esq., President of the Academy. These latter were put in glycerine only, and are in better condition than those received by Dr. Blake.

Of the specimens received from Mr. Lawson, some individuals are younger than either of Dr. Blake's. In these the polyp-rows are farther apart, and there are not so many polyps in the row; neither do the ends of the rows approximate so closely on the side opposite the axial edge; the polyps being not nearly so many in the same length, or presenting (as do some of Dr. Blake's specimens) so crowded an appearance. In cross-section through the polypiferous portions, the younger individuals are less oval or acutely-ovate than in the older specimens. A comparison of individuals indicates an external differentiation, analagous to that displayed by specimens of the same species in Virgularia. The general aspect of this species, judging from the figure in Plate XXXI. of Johnston's British Zoophytes (2d ed.), is like *P. quadrangularis* from Oban, only in that species the rows of polyps it is stated, are composed of "four, five or six polyps in a row," one figure showing seven.

Description of a New Genus and Species of Alcyonoid Polyp.*

BY ROBERT E. C. STEARNS.

At a meeting of the California Academy of Sciences, held on the third day of February, 1873, a paper was read by me, entitled "Remarks on a New Alcyonoid Polyp, from Burrard's Inlet;"† in which I gave a *resumé* of the discussions, notices, etc., in this country and in England, arising from the examination by several naturalists, of certain "switch"-like forms, which had been received by different parties from the Gulf of Georgia (more particularly from Burrard's Inlet, in said gulf); several specimens of said "switches" being in the Museum of the California Academy.

These "switches," or rods, were referred by Dr. Gray, of the British Museum, to his genus "Osteocella," and by Mr. Sclater's correspondent stated to belong to "a sort of fish"; but by the majority of scientific gentlemen who had seen these "switches" they were regarded as belonging to a species of Alcyonoid Polyp. I expressed the belief that they belonged to a species of *Umbellularia*.

At a meeting of the California Academy, held on the evening of August 4, 1873, Dr. James Blake presented a specimen of the polyp of which these so-called switches are the axes, which had been sent to him from the Gulf of Georgia by his friend, Capt. Doane. This specimen was one of six or seven sent at the same time, all of which were in a tolerable state of preservation, though, as might have been anticipated, the more delicate tissues of the polyps are somewhat decomposed, and some of the specimens are in some places lacerated. They all are, however, sufficiently perfect to determine the true position, and show that the "switches" are, as was supposed, the supporting stalks or axes of an Alcyonoid Polyp "related or pertaining to the group *Pennatulidæ*."

At the last meeting I referred the specimen before the Academy to that division of the *Pennatulidæ* known as *Virgularia*, but upon a subsequent examination of the authorities, I find that those forms in which the axis is unilateral, or on one side, come within the Genus *Pavonaria* of Cuvier.

The only species heretofore described so far as I can learn, and on which this genus is based, is *P. quadrangularis*, of which a lengthy and interesting description from Prof. Forbes, is given in Johnston's British Zoöphytes (Vol. I, pp. 164–166). In that species, however, the axis is "acutely quadrangular," and the polyps are arranged in three longitudinal series, corresponding to three of the "angles of the stem."

---

* Printed in advance August 20th, 1873.
† *Vide* Proc. Cal. Acad. Sciences, vol. V, part I, pp. 7–12.

In the specimen presented by Dr. Blake the style or axis is round, and the polyps are arranged in two longitudinal unilateral series, which conform to the convexity of the external fleshy covering. With these differences, I think I am justified in placing it in a new sub-genus for which I propose the name of *Verrillia*, in honor of Prof. Verrill of Yale College.

Genus PAVONARIA, Cuvier.

Sub-genus VERRILLIA, Stearns.

Polypidom linear-elongate, round, oval or ovate in cross-section. Axis round, slender, bony; polyps arranged in two unilateral longitudinal series.

Verrillia Blakei, Stearns; n. s.

Polyp-mass or polypidom, of a flesh or pink color, linear, elongate, attenuate; polypiferous portion about three fourths of the entire length, rounded oval to ovate-elliptic in cross section, and from three fourths to one inch in greatest diameter, flatly tapering toward the tip, as well as decreasing in the opposite direction to where the polypiferous rows terminate or become obsolete. From this latter point to the beginning of the base or root, a portion of the polypidom, equal to about one sixth of its entire length, is quite slender, being only about twice the diameter of the naked axis, and the surface quite smooth; said portion, as well as the base, is round (in cross section); the basal part is from one ninth to one eleventh of the entire length, and about one inch in diameter, with the surface longitudinally wrinkled or contracted, presenting a ridged or fibrous appearance.

Style or axis long, slender, white, hard, bony, somewhat polished, about three sixteenths (3-16) of an inch in diameter in the thickest part, tapering gradually toward the tip, and attenuated, with surface somewhat roughened toward the basal extremity. Enclosed in the polyp-mass or polypidom, the axis is central from the base to where the polyp-rows begin, when it soon becomes marginal or lateral, forming a prominent rounded edge (free from polyps) on one side of the polypiferous portion of the whole.

From near the sides of the axial edge the polyp-rows start, and run obliquely upward to the opposite side, where they nearly meet, presenting, when that side is observed from above, a concentric chevron or $\wedge$-like arrangement, modified by the convexity of the polypidom. The more conspicuous polyp-rows show from nine to fourteen polyps, with occasional intermediate rows of three or more polyps.

The length of the most perfect of Dr. Blake's specimens was sixty-six (66) inches; of which, commencing at the tip, a length of forty-eight and a quarter ($48\frac{1}{4}$) inches was occupied by the polyp-rows, which numbered two hundred and forty-five (245), or twice that number when both sides or arms of the chevron or $\wedge$ are considered. The number of polyps in each row was, in this specimen, from eight (8) to eleven (11), with occasional intermediate shorter rows of from three (3) to seven (7). Estimating ten to the [...], this specimen exhibited about *five thousand* polyps, all of which, as well as the polyps in the other specimens, were filled with ova, of an orange color. In the next section of this specimen, the length between the last polyp-row and the swell of the base or root, is eleven and one quarter (11¼) inches; thence to the termination of the base, six (6) inches.

In some specimens, the polypiferous portion makes from one to two turns around the axis in its entire length. Plate IX, fig. 1, exhibits the general aspect of the species, reduced to a scale of one inch to the foot; fig. 2, a section of the polypiferous part of one of the oldest and largest specimens.

The average dimensions of thirty-six (36) of the axes in the Museum of the California Academy is five feet six and one third inches in length, and the diameter of the largest, nine thirty-seconds of an inch; diameter of smallest specimen, one sixteenth of an inch.

Dr. Blake's specimens were preserved in a mixture of glycerine and alcohol, and the more delicate tissue of the polyps appears to have been somewhat injured by the latter ingredient.

Additional specimens of the above species, from the same locality, have been received from J. S. Lawson, Esq.,* of the U. S. Coast Survey, by George Davidson, Esq., President of the Academy. These latter were put in glycerine only, and are in better condition than those received by Dr. Blake. Of the specimens received from Mr. Lawson, some individuals are younger than either of Dr. Blake's. In these the polyp-rows are farther apart, and there are not so many polyps in the row; neither do the ends of the rows approximate so closely on the side opposite the axial edge; the polyps being not nearly so many in the same length, or presenting (as do some of Dr. Blake's specimens) so crowded an appearance. In cross-section through the polypiferous portions, the younger individuals are less oval or acutely-ovate than in the older specimens. A comparison of individuals indicates an external differentiation, analogous to that displayed by specimens of the same species in Virgularia. The general aspect of this species, judging from the figure in Plate XXXI of Johnston's British Zoöphytes (2d ed.), is like *P. quadrangularis* from Oban, only in that species the rows of polyps, it is stated, are composed of "four, five or six polyps in a row," one figure showing seven.

I have named this species for Dr. James Blake, of San Francisco, author of many valuable scientific papers, to whom I am indebted for numerous courtesies.

---

*The following remarks accompanied the specimens received from Mr. Lawson: "Obtained from fishermen by J. C. Hughes, Esq., of Burrard's Inlet, (Gulf of Georgia), at the request of Jas. S. Drummond, Esq., of Victoria, who kindly and keenly interested himself for me. JAMES S. LAWSON."

INDEX TO PLATE IX.

FIG. 1.—Verrillia Blakei; Stearns General aspect; one-twelfth natural size; from Dr. Blake's specimens.

FIG. 2.—Section of Polypiferous portion of one of the largest and most crowded specimens. Natural size.

FIG. 3.—Cross-section through Polypiferous part; a, principal longitudinal canal; b, axis.

FIG. 4.—Cross-section through basal part; a, canal; b, axis.

FIG. 5.—Section of Polypiferous portion of a smaller and less crowded specimen, received from J. S. Lawson, Esq. Natural size.

FIG. 6.—Section of above; (Fig. 5) showing chevron-like arrangement of Polyp-rows, opposite the axial-side. Natural size.

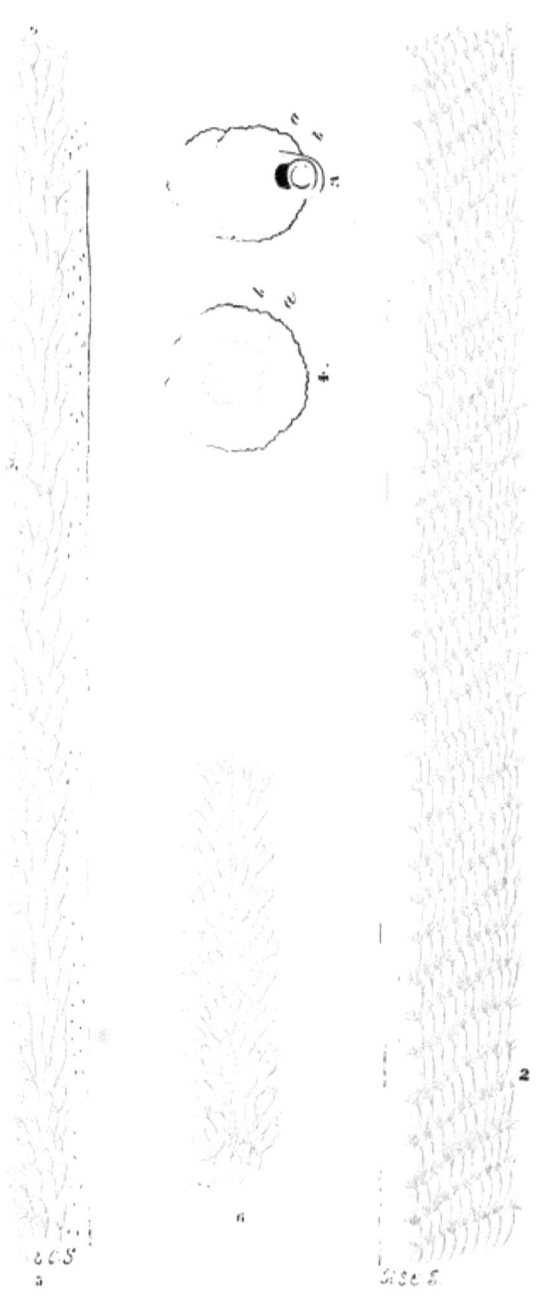

VERRILLIA BLAKEI, STEARNS.
(SEE PAGE 18.)

Remarks Suggested by Dr. J. E. Gray's Paper on the "Stick Fish," in "Nature," Nov. 6th, 1873.

BY ROBERT E. C. STEARNS.

At a meeting of this Academy on the 3d of February, 1873, certain switch-like rods, being the axes of some polyp-form, as well as the general characters of Alcyonoid Polyps, were considered and discussed, for the purpose of tracing by analogy and determining the relations and position of the specimens under consideration at that time; and it may be remembered that a paper was read, in which was given at considerable length a resume of what had appeared in the columns of *Nature*, in the way of notes and comments by several learned gentlemen.

These rods, switches, or wands, as the specimens had been variously called, were first brought to the notice of the Academy on the 5th of June, 1871, when specimens were presented to the Museum, and, so far as an opinion was expressed at that time in a general way, the specimens were placed near the group to which it has been subsequently proved that they belong.

On the 4th of August, 1873, Dr. James Blake submitted an entire specimen of the polyps, of which the rods, etc., are the central stalks or axes: that is, one of these rods or switches was presented by him, with the investing soft or fleshy covering, which proved it to be either a *Pavonaria*, or closely related to that genus. Accordingly, I published a description placing it in the genus *Pavonaria*, and gave it the specific name of "Blakei," (*Pavonaria Blakei*) and the same was printed in the *Mining and Scientific Press* of this city, August 9th, 1873.

Before the succeeding regular meeting of the Academy, which took place August 18th, 1873, through access to more recent literature bearing on the subject, I perceived at once that not only was the species new, but that its separation generically was warranted, and the sub-genus *Verrillia* was made by me to receive it; and a description of the genus and species was read at that meeting, and printed copies of my paper (dated August 20th) were sent to various authors, societies, and scientific journals, in advance of the regular publication of the Academy's Proceedings.

Among the many scientific gentlemen who had discussed the character and relations of the so-called switches, Dr. P. L. Sclater, of the Zoölogical Society, kindly gave publicity to *Verrillia Blakei*, in *Nature*, for October 9th, 1873.

In the same journal, of date Nov. 6th, 1873, Dr. J. E Gray, of the British Museum, publishes a communication "On the stick-fish, (*Osteocella septentrionalis*) and on the habits of sea-pens," in which he refers to a specimen presented to the Museum by Mr. Coote M. Chambers, and of which he says: "Unfortunately the specimen did not arrive in a good state for exhibition. The greater part of the animal portion had been washed off, probably by the motion of the solution during the transit; only about a foot of the flesh which was loose on the axis, and the thick, swollen, naked, club-shaped base, without polypes, remained; but it was in a sufficiently good state to afford the means of determining its zoölogical situation, and of examining its microscopical and other zoölogical characters."

In the next paragraph, of which I quote a portion, Dr. Gray says: "Mr. Chambers' specimen is the animal of the axis or stick, that I described as *Osteocella septentrionalis*, (Ann. and Mag. Nat. Hist., 1872, p. 406) * * * * * and is evidently the same animal as *Pavonaria Blakei*, described by R. E. C. Stearns."

"Two days after I received this specimen, I received by post Mr. Stearns' description of the stick fish, (*Pavonaria Blakei*) from the San Francisco *Mining and Scientific Press*, August 9th, 1873."

Towards the close of his article, Dr. Gray writes: "Mr. Stearns' paper, in the Proceedings of the California Academy of Sciences, is a reprint of the paper in the San Francisco *Mining and Scientific Press*, with a few additions, and the addition of a new sub-genus, *Verrillia*, although he quotes *Osteocella*." In this paper Mr. Gray gives what he considers "the synonymy of those animals"; first, the genera, and next, the species; placing my first generic determination. *Pavonaria*, and my subsequent sub-genus, *Verrillia*, in the order as recited, as synonymes of his genus *Osteocella*.

I would ask Dr. Gray by what warrant, either of science or justice, he places *Pavonaria* or *Verrillia*, definitely described genera, as synonymes of his indefinite and vague *Osteocella*, which latter he *publishes* as a genus, for it cannot be said he *describes* it, in the "Catalogue of Sea-Pens—or Pennatulidæ—in the British Museum" 1870, page 40. Gray's genus *Osteocella* is based upon a "bone," (probably the axis of a polyp) which was sent to the British Museum "many years ago," from Australia, by a gentleman named Clifton. The investing fleshy substance, or soft portion of the animal, of which said bone formed a part, had not been seen by Dr. Gray at the time he invented the name *Osteocella*, and even to this date no additional light has been furnished by him regarding the Australian form. He was not even *positive* that the "bone" belonged to a zoöphyte, for he says: "or, it may be the long conical bone of a form of decapod cephalopod which has not yet occurred to naturalists, as Mr. Clifton spoke of its being a free marine animal: it has a cartilaginous apex like the cuttle fish."

In which of the great divisions of the animal kingdom does Dr. Gray place it, or did he place his Australian bone in 1870?

Courtesy and fairness suggest that as he printed it in the Catalogue of *Pennatulidæ*, it should be conceded, as I have written, in a previous paper, "that, in his mind, the balance of reasoning tends in that direction."

Admitting this latter, what then? The Australian bone upon which rests his genus *Osteocella* is described by Dr. Gray as being "thick, about eleven inches long, tapering at each end." Subsequently he has received one of the stalks, or axes, of what I have named *Verrillia Blakei*; of the latter, he says it is "long, slender, about sixty-four inches long, attenuated at the base, and very much attenuated and elongated at the other end." "Mr. Carter" examined both of the bones referred to, microscopically, and "finds them" to "present the same horny structure," etc. An examination with acid was made, but as it would be rather difficult to comprehend in what way generic or specific determinations within any related groups could be determined by *acid*, this test may be allowed to pass.

The reference of *Verrillia* to *Osteocella* as a synonyme, or otherwise, must rest on this microscopic test, as the soft investing portion of the animal, the perfect or complete polyp or polypidom of the Australian form, to which the bone, if the axis of an alcyonoid, belongs, and upon which Dr. Gray made his genus *Osteocella*, has not, as yet, been seen by him, or brought under scientific observation. He cannot aver, because he does not know, but that it may be a species which belongs to some genus already described, or that it may properly fall in as a sub-genus of some of the genera of Alcyonoids previously known; he does not know but what its relationship may be nearer to any of the other groups than to *Pavonaria*. No description sufficiently accurate to be worthy of consideration can be made from the axial rods or bones alone, of this class of animal forms, nor can species be satisfactorily determined without the fleshy portion; nor, in the present state of our knowledge, can the microscope determine these points.

In his genus *Osteocella*, which, it must be borne in mind, rests solely on the *naked* Australian bone or axis, which he says is "thick," "eleven inches long," as published in the British Museum Catalogue of *Pennatulidæ*, no information is furnished as to the soft investing portion, for the very good reason that it had not been seen by him; yet in the number of *Nature* last quoted, he speaks of "the complete polyp-mass," thus clothing his west Australian *Osteocella* with the fleshy covering of the west North-American *Verrillia*. So much for his generic synonymy. As to the species, the North-American form, as referred to by him, could not be definitely placed, by anything written by Dr. Gray prior to the date of my description.

This is a matter, not of personal pride, but of scientific accuracy; and scientific naturalists should not lose sight of, or be diverted from, this *sine qua non*, or palliate individual idiosyncracies which involve integrity, and which should not be allowed to pass without challenge or comment.

Aboriginal Shell Money.

BY ROBERT E. C. STEARNS.

Of the numerous objects or substances which exist in a natural state, and which require little or no mechanical preparation for adaption for use as money, the shells of many of the marine *mollusca*—or shell-fish, so called—furnish at once an excellent and appropriate material. Where the metals do not exist, or the knowledge of manipulating them is wanting, no substance or form can be named which is at once so available and convenient. Thus we find that certain forms of shells have been used by the aborigines of both shores of our own continent; and, though the forms used by the Indians of the Atlantic Coast were quite different, according to the authors whom we have consulted, from that of the money of the West American tribes, yet this can not be accounted for on the supposition that a similar form is not found on the Atlantic Coast, for such is not the fact. It is not unreasonable to suppose that they had but little, if any, knowledge of each other, and more likely none at all. Being separated by the breadth of a continent, with many wide and rapid rivers and several lofty mountain ranges intervening, and the intermediate country occupied by numerous and distinct tribes quite as jealous of any invasion of their territory as are the civilized nations of to-day, the use or the knowledge of the use of any substance or particular form for money by the tribes of either coast, was probably unknown to those of the opposite trans-continental shore.

The Pilgrim settlers of the Massachusetts Colony at Plymouth found a form of money in use among the Indians of New England; and in the Historical Collections of Massachusetts, and from other sources as recorded by Governor Winthrop and Roger Williams, we are informed as to its character and substance. One of the most common bivalve mollusks (clams) of that coast is the *Venus mercenaria*, or *Mercenaria violacea*, (Plate VI, fig. 1,) as it is now called by naturalists; it is the "hard-shell clam" of the New York market, and in the markets of Boston is known as the "quahog." The valves or shells of this species frequently display an interior purple edge—varying in this respect, it is said, in different localities—the rest of the shell being of a clear white. From the darker colored portion the Indians made their purple money, or *wampum*, as it was called; while from the axis of a species of *Pyrula* or conch, and from

---

* See also *Overland Monthly* for October, 1873.

other shells, they made their white money, or white *wampum*. In reference to the first shell, and its use as a substance from which the wampum was made, we have the following: "The quahaug (*Venus mercenaria*), called by Roger Williams the *poquau* and the *hen*, is a round, thick shell-fish, or, to speak more properly, worm. It does not bury itself but a little way in the sand; is generally found lying on it, in deep water; and is gathered by rakes made for the purpose. After the tide ebbs away, a few are picked up on the shore below high-water mark. The quahaug is not much inferior in relish to the oyster, but is less digestible. It is not eaten raw; but is cooked in various modes, being roasted in the shell, or opened and broiled, fried, or made into soups and pies. About half an inch of the inside of the shell is of a purple color. This the Indians broke off and converted into beads, named by them *suckauhock*, or black money, which was twice the value of their *wampom*, or white money, made of the *metauhock*, or periwinkle (*Pyrula*).*

"As to the derivation of the word 'quahog,' Governor Winthrop refers to it as '*poqua'uuges*, a rare shell and dainty food with the Indians. The flesh eats like veal; the English make pyes thereof; and of the shells the Indians make money.' He says of the money, 'It is called *Wampampeege*. † Also, called by some English *hens-po-qua-hock*; three are equal to a penny; a fathom is worth five shillings.' ‡

"*Poquahock*, corrupted into *quahaug* or *quahog*."

The money or *wampum* made from the shells above referred to, was not only in use among the Indians, but among the Whites also. Col. T. W. Higginson, of Massachusetts, in one of his *Atlantic Essays*, "The Puritan Minister," says: "In coming to the private affairs of the Puritan divines, it is humiliating to find that anxieties about salary are of no modern origin. The highest compensation I can find recorded, is that of John Higginson, in 1671, who had £160 voted him in 'country produce,' which he was glad, however, to exchange for £120 in solid cash. Solid cash included beaver-skins, black and white *wampum*, beads and musket-balls, value one farthing."

In Cadwalader Colden's *History of the Five Indian Nations* (p. 34), he says that wampum is made of the large whelk-shell *Buccinum*, and shaped like long beads: it is the current money of the Indians. Whether the shells of the true *Buccinum* (*B. undatum*, Linn., or *B. undulatum*, Mull.), or those of *Busycon* (*B. canaliculatum* and *B. carica*), is not satisfactorily explained.

In Major Rogers' *Account of North America* (London 1765), in alluding to the *wampum* of the Indians, he says: "When they solicit the alliance, offensive, or defensive, of a whole nation, they send an embassy with a large belt of wampum and a bloody hatchet, inviting them to come and drink the blood of their enemies. The *wampum* made use of on these and other occasions, before their acquaintance with the Europeans, was nothing but small shells, which they picked

---

* Massachusetts Historical Society's Collections, VIII, 192 (1802).

† Journal Royal Society, June 27, 1634.

‡ Vide *Invertebrata of Massachusetts*, Binney's edition, p. 134.

up by the sea-coast, and on the banks of the lakes; and now it is nothing but a kind of cylindrical beads, made of shells, white and black, which are esteemed among them as silver and gold are among us. They have the art of stringing, twisting, and interweaving them into their belts, collars, blankets, moccasins, etc., in ten thousand different sizes, forms, and figures, so as to be ornaments for every part of dress, and expressive to them of all their important transactions.

"They dye the *wampum* of various colors and shades, and mix and dispose them with great ingenuity and order, so as to be significant among themselves of almost everything they please; so that by these, their words are kept and their thoughts communicated to one another, as ours by writing. The belts that pass from one nation to another in all treaties, declarations, and important transactions, are very carefully preserved in the cabins of their chiefs, and serve not only as a kind of record or history, but as a public treasure."

Colden is the only author in whose writings we find any allusion to the use or manufacture of money or *wampum* by any of the *interior* tribes, and the tribes of the Five Nations were not remote from the Atlantic shore.

How far to the south of New England this *wampum* money was used, we do not know. The shells of which it was made are abundant in the neighborhood of New York and Philadelphia, and are the common clam in the markets of those cities. A closely related form (*Mercenaria prapacca*, Say), is found on the shores of Florida, and attains an exceedingly large size; specimens collected in Tampa Bay frequently weigh as much as three and a half pounds after the animal is removed. Explorations made by us in that State in the year 1869, in the course of which many of the ancient shell-heaps and burial-mounds on both shores of the peninsula were examined, resulted in the obtainment of much interesting material, but no specimens were found of forms which suggested their possible use for money.

Crossing the continent to the north-western coast of North America, we find that the sea-board aborigines had, and in a decreasing degree still use, a money of their own—a species of shell, though of a widely different form from that used by the natives of the Atlantic coast. The money of the West-coast Indians is a species of tusk-shell (*Dentalium*), resembling in miniature the tusks of an elephant. (Plate VI, fig 2). Mr. J. K. Lord, formerly connected, as naturalist, with the British North American Boundary Commission, refers to the use of these shells as money "by the native tribes inhabiting Vancouver's Island, Queen Charlotte's Island, and the main-land coast from the Straits of Fuca to Sitka. Since the introduction of blankets by the Hudson's Bay Company, the use of these shells has to a great extent died out; and the blankets have become the money, as it were, by which everything is now reckoned and paid for by the savage. A slave, a canoe, or a squaw, is worth in these days so many blankets; it used to be so many strings of *Dentalia*." Mr. Lord also remarks: "The value of the *Dentalium* depends upon its length. Those representing the greater value are called, when strung together end to end, a *Hi-qua*; but the standard by which the *Dentalium* is calculated to be fit for a *Hi-qua* is that twenty-five shells placed end to end must make a fathom, or six feet in length. At one time a *Hi-qua* would purchase a male slave, equal in value to fifty blankets, or £50 sterling.*

Mr. Frederick Whymper, speaking of an Indian muster of various tribes at or near Fort Yukon, Alaska, in 1867, says: "Their clothing was much be-fringed with beads, and many of them wore through the nose (as did most of the other Indian *men* present) an ornament composed of the *Hya-qua* shell (*Dentalium entalis*, or *Entalis vulgaris*). Both of the fur companies on the river trade with them, and at very high prices. These shells were formerly used, and still are, to some extent, as a medium of currency by the natives of Vancouver Island and other parts of the North-west Coast. I saw on the Yukon, fringes and head-ornaments, which represented a value in trade of a couple of hundred marten-skins.† Mr. Whymper further remarks that "These shells are generally obtained from the west coast of Vancouver Island," and that his spelling "*Hya-qua* conveys a "closer approximation to the usual pronunciation of the word" than Mr. Lord's "*Hi-qua*."

The use of these shells for nasal ornamentation by the Indians, as observed by Mr. Whymper at Fort Yukon, attracted our attention while at Crescent City, in this State, in the year 1861. A medicine-man, belonging to one of the neighboring tribes, had perforated the partition which separates the nostrils, and, into the hole thus made, had inserted from each side, point by point, two of these shells, which decoration was further increased by sticking a feather of some wild-fowl into the large end of each of the hollow shells.

As to the length of the shells, as implied by Mr. Lord's statement "that twenty-five shells placed end to end must make a fathom or six feet," we are inclined to believe there is some mistake, as the shells would have to average very nearly three inches in length. Of the great number which we have seen of the species mentioned by Lord and Whymper (*Dentalium entalis*, or *Entalis vulgaris*), but very few attain a length of two inches; the great majority averaging much less. As to the specific names of the shells used as above, and the localities from which they are obtained, it may be well to state that the "west coast of Vancouver Island" form is the *Dentalium Indianorum*‡ of Dr. P. P. Carpenter; but probably the greater part of the tusk-shells which are or have been in circulation, do not belong to the American species, but to the common European *Dentalium*,§ referred to by the gentleman, and which closely resembles the American. The foreign species has been extensively imported for the Indian trade, and we have noticed at different times large numbers of the imported shells displayed for sale in the fancy goods stores in San Francisco, together with beads and other Indian goods. The use of the *Dentalia* for money among the Alaskan tribes is also corroborated by Mr. W. H. Dall, whose extensive travels and thorough investigations in that territory are well known. It

---

* Proceedings Zoological Society, London, March 8th, 1864.

† Whymper's *Alaska*, Harper's edition, 1869, p. 255.

‡ Supp. Rep. Brit. Ass'n, 1863. on Mollusca of W. N. America, p. 648.

§ *Antalis entalis*. *Vide* Adams' *Genera*, vol. I, p. 457.

is highly probable that the use of these shells in that region will soon become a story of the past, and the money of the Pale-faces will supersede among the Red-men the shells of the sea.

The Indians of California, or the tribes inhabiting the northern portion of the coast and the adjoining region, also use the tusk-shells for money; either the shells or the shell-money is called *alli-co-check*, or *allicochick*—the latter being the orthography, according to Mr. Stephen Powers, whose valuable papers upon "The Northern California Indians," in the *Overland Monthly*, are an important contribution to American aboriginal history.

"For money, the Cahrocs make use of the red scalps of woodpeckers, which are valued at $5 each; and of a curious kind of shell, resembling a cock's spurs in size and shape, white and hollow, which they polish and arrange on strings, the shortest being worth twenty-five cents, the longest about $2—the value increasing in a geometrical ratio with the length. The unit of currency is a string the length of a man's arm, with a certain number of the longer shells below the elbow, and a certain number of the shorter ones above. This shell-money is called *allicochick*, not only on the Klamath, but from Crescent City to Eel River, though the tribes using it speak several different languages. When the Americans first arrived in the country, an Indian would give from $40 to $50 in gold for a string of it; but now it is principally the old Indians who value it at all." *

In speaking about marriage among the Eurocs, he says: "When a young Indian becomes enamored of a maiden, and cannot wait to collect the amount of bells demanded by her father, he is sometimes allowed to pay half the amount, and become what is termed 'half married.' Instead of bringing her to his cabin and making her his slave, he goes to live in *her* cabin and becomes *her* slave." Again, he says: "Since the advent of the Americans, the honorable estate of matrimony has fallen sadly into desuetude among the young braves, because they seldom have shell-money now-a-days, and the old Indians prefer that in exchange for their daughters.... (The old generation dislike the white man's money, but hoard up shell-money like true misers)," etc. Among the Hoopas, "murder is generally compounded for by the payment of shell-money." †

In connection with the use of money in traffic among the interior Indians, it appears that "all the dwellers on the plains, and as far up on the mountain as the cedar-line, bought all their bows and most of their arrows from the upper mountaineers. An Indian is about ten days in making a bow, and it costs $3, $4, or $5, according to the workmanship; an arrow, 12½ cents. Three kinds of money were employed in this traffic. White shell beads, or rather buttons, pierced in the centre and strung together, were rated at $5 per yard; periwinkles, at $1 a yard; fancy marine shells at various prices from $3 to $10, or $15, according to their beauty." ‡

---

* *Vide Overland Monthly*, vol. VIII, pp. 329, 427, 535.

† *Id.*, vol. IX, p. 156.

‡ *Id.*, vol. X, p. 325.

The shell-money here referred to is not sufficiently particularized to admit of a determination of the species to which the shells belonged. In connection with the treatment of the sick among the Meewocs, Mr. Powers says: "The physician's prerogative is, that he must always be paid in advance; hence, a man seeking his services brings his offering along—a fresh-slain deer, or so many yards of shells, or something—and flings it down before him without a word, thus intimating that he desires the worth of that in medicine and treatment. The patient's prerogative is, that if he dies, his friends may kill the doctor."*

Among the Moi-locs, or Modocs, "when a maiden arrives at womanhood, her father makes a kind of a party in her honor. Her young companions assemble, and together they dance and sing wild, dithyrambic roundelays, improvised songs of the woods and the waters:

> "'Jumping echoes of the rock;
> Squirrels turning somersaults;
> Green leaves, dancing in the air;
> Fishes white as money-shells,
> Running in the water, green, and deep, and still.
> Hi-ho, hi-ho, hi-hay!
> Hi-ho, hi-ho, hi-hay!'

This is the substance of one of the songs, as translated for me." †

Among the Yocuts, another California tribe, whose dominion covers "the Kern and Tulare basins, and the middle San Joaquin," etc., "their money consists of the usual shell-buttons, and a string of them reaching from the point of the middle finger to the elbow is valued at twenty-five cents. A section of bone, very white and polished, about two and a half inches long, is sometimes strung on the string, and rates at a 'bit.' They always undervalue articles which they procure from Americans. For instance, goods which cost them at the store $5, they sell among themselves for $3." ‡

We have no authentic *data* as to whether the value of the shell-money, properly so-called, among the California Indians, and those farther north, was graduated by the color, or whether they generally used other than the *hya-qua* or *allicochick* (*Dentalia*), which are white and have a shining surface; for though, as above, "periwinkles" and "fancy marine-shells" are mentioned as used in trade, these may have been regarded more as articles of ornamentation, and esteemed among the interior Indians particularly as precious, the same as diamonds and fine jewelry are among civilized people. In this view, the interior Indians of California are probably not unlike the more southern Indians of New Mexico, for a friend of ours (Dr. Edward Palmer of the Smithsonian Institution) informed us a few years ago, that while traveling in that territory he was witness to a trade wherein a horse was purchased of one Indian by another, the price paid being a single specimen of the pearly ear-shell (*Haliotis rufescens*), or common California redback *abalone* or *avlon*.

---

* *Overland Monthly*, vol. X, p. 327.

† *Id*., p. 541.

‡ *Id*., vol. II, p. 108.

As to the value of the tusk-shells among the California Indians, the method of reckoning the same is by measuring the shells on the finger-joints, the longest being worth the most.

We have been informed that the Indians who formerly resided in the neighborhood of the old Russian settlement of Bodega, used pieces of a (bivalve) clam-shell (*Saxidomus aratus**) for money, but we have been unable to obtain a specimen, or to verify the statement. Recently, our friend Mr. Harford, of the Coast Survey, has discovered in some Indian graves, on one of the islands off the southerly coast of this State, beads, or money, of a different character from any heretofore observed. These were made by grinding off the spire and lower portion of a species of univalve shell (*Olivella biplicata*, Sby., Plate VI, Fig. 3), so as to form small, flat, button-shaped disks with a single central hole. These much resemble in form some of the *wampum* of the New England tribes. Another variety was found in the same places by the gentleman named, which was made of a species of key-hole limpet-shell (*Lucapina crenulata*, Sby., Plate VI. Fig. 6), of much larger size than that first mentioned. So far, however, as we have investigated, these last described forms of shell-money are not in use among the California Indians of the present day. Plate VI. Figures 6ª and 6ᵇ represent beads or money made from *Lucapina*.

The use of shells for money is not peculiar to the natives of North America. The well-known and exceedingly common money cowry (*Cypraea moneta*, Plate VI. Figs. 5, and 5ª) or "prop-shell," an inhabitant of the Indo-Pacific waters, "is used as money in Hindostan and many parts of Africa.... Many tons are ....imported to.... Great Britain and.... exported for barter with the native tribes of western Africa." †

Reeve mentions in the second volume of the *Conchologia Systematica*, that "a gentleman residing at Cuttack, is said to have paid for the erection of his *bungalow* entirely in these *cowries* (*C. moneta*). The building cost him about 4,000 *rupees sicca* (£400 sterling), and, as sixty-four of these shells are equivalent in value to one *pice*, and sixty-four *pice* to a *rupee sicca*, he paid for it with over 16,000,000 of these shells."

Though the number above mentioned is very large, the prop-shell is an exceedingly abundant form. We have received in a single box from the East Indies not less than 10,000 specimens at one time. "In the year 1848, sixty tons were imported into Liverpool, and in 1849, nearly three hundred tons were brought to the same port."

The following extract from a paper by Prof. E. S. Holden, on *Early Hindoo Mathematics*,‡ justifies the inference that the use of the *Cypraea moneta* for money has a very considerable antiquity, and quite likely extends back to a period many centuries earlier than the date of the treatise.§ "The treatise continues rap-

---

* S. aratus+S. gracilis, Gld.

† Baird's *Dictionary of Natural History*, p. 193.

‡ *Popular Science Monthly*, July, 1873, p. 337.

§ "This treatise, the *Lilivati of Bhascara Acharya*, is supposed to have been a compilation, and there are reasons for believing a portion of it to have been written about A. D. 628. However this may be, it is of the greatest interest, and its date is sufficiently remote to give to Hindoo mathematics a respectable antiquity."

idly through the usual rules, but pauses at the reduction of fractions to hold up the avaricious man to scorn: 'The quarter of a sixteenth of the fifth of three-quarters of two-thirds of a moiety of a *drumme* was given to a beggar by a person from whom he asked alms; tell me how many cowry shells the miser gave, if thou be conversant in arithmetic with the reduction termed subdivision of fractions.'" These shells are also known as "Guinea money," and, it is said, have been used as a financial medium in connection with the African slave-trade. Doubtless many a poor negro has been sold, and has lost his liberty, for a greater or less number of these shells.

Another species of cowry of small size, and which inhabits the Indo-Pacific province, called the "ringed cowry" (*Cypræa annulus*), the back or top of the shell being ornamented with an orange-colored ring, "is used by the Asiatic islanders to adorn their dress, to weight their fishing-nets, and for barter. Specimens of it were found by Dr. Layard in the ruins of Nimroud."

According to the relation of a recent voyage, transactions are performed in Soudan by barter, or by means of small shells picked up in the Niger, which are called *oudâas* or *woodahs*.[*]

It will be seen, therefore, that shells have been and are still used as money by portions of the human race, but to an extent much less than formerly. It would be quite difficult to point out any other natural production which is more appropriate, when size, shape and substance are considered.

---

[*] *Science Gossip*, Dec., 1866, p. 283.

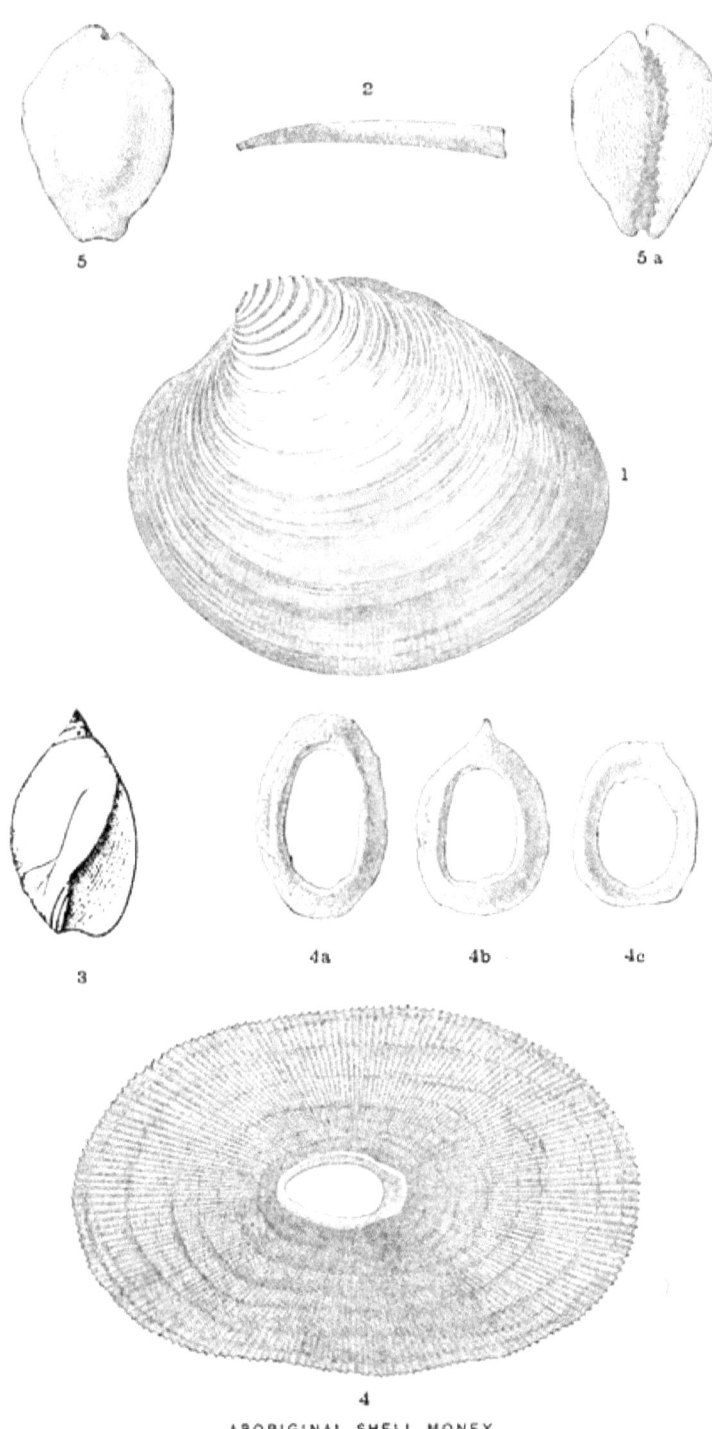

ABORIGINAL SHELL MONEY.
[SEE PAGE 113.]

Tellina Cumingii, *Hanl.*
Gulf of California.

Strigilla carnaria, *Linn.*
San Juan del Sur, Nicaragua.

Donax assimilis, *Hanl.*
Panama.

Trigona radiata, *Sby.*
Carmen Island, Gulf of California.

Dosinia Dunkeri, *Phil.*
Panama.

Dosinia ponderosa, *Gray.*
Carmen Island, Gulf of California.

Callista aurantia, *Hanl.*
Gulf of California.

Callista chionaea, *Mke.*
Carmen Island, Gulf of California.

Tapes histrionica, *Sby.*
Panama.

Cardita affinis, *Brod.*
Panama.

Cardita radiata, *Brod.*
Panama.

Codakia tigerina, *Linn.*
Gulf of California.

Lucina muricata, *Chemn.*
Gulf of California.

Lucina eburnea, *Rve.*
Gulf of California.

Lucina undata, *Cpr.*
Gulf of California.

Perna Chemnitzianus, *D'Orb.*
Panama.

Janira dentata, *Sby.*
Gulf of California.

Bulla Adamsi, *Mke.*
Gulf of California.

Siphonaria lecanium, *Phil.*
Gulf of California.

Siphonaria pica, *Sby.*
Gulf of California.

Dentalium semipolitum, *Cpr.*
Gulf of California.

Callopoma fluctuosum, *Mawe.*
Panama.

Callopoma saxosum, *Wood.*
Panama.

Uvanilla inermis, *Gmel.*
Panama.

Calliostoma versicolor, *Mke.*
Gulf of California.

Tegula pellis-serpentis, *Wood.*
Panama.

Omphalius viridulus, *Gmel.*
Panama.

Nerita Bernhardi, *Rcl.*
Panama.

Nerita scabricosta, *Lam.*
Panama.

Neritina intermedia, *Sby.*
Nicaragua.

Neritina latissima, *Brod.*
Panama.

Neritina picta, *Sby.*
Panama.

Cerithium famelicum, *C. B. Ad.*
Acapulco.

Cerithium maculosum, *Kien.*
Panama.

Cerithium stercus-muscarum,
Panama. [*Val.*

Vertagus gemmatus, *Hds.*
Acapulco.

Cerithidea Montagnei, *D'Orb.*
Panama.

Cerithidea pulchra, *C. B. Ad.*
Panama.

Cerithidea varicosa, *Sby.*
Panama.

Cerithidea Mazatlanica, *Cpr.*
Gulf of California.

Littorina aspera, *Phil.*
Nicaragua.

Littorina conspersa, *Phil.*
Nicaragua.

Littorina varia, *Sby.*
Panama.

Littorina fasciata, *Gray.*
Panama.

Littorina pulchra, *Phil.*
Panama.

Modulus catenulatus, *Phil.*
Gulf of California.

Modulus disculus, *Phil.*
Gulf of California.

Planaxis planicostata, *Sby.*
Panama.

Radius variabilis, *C. B. Ad.*
Gulf of California.

Cypraea exanthema, *Linn.*
Panama.

Cypraea arabicula, *Lam.*
Gulf of California.

Cypraea punctulata, *Gray.*
Panama.

Cypraea Sowerbyi, *Kien.*
Gulf of California.

Cypraea albuginosa, *Mawe.*
Gulf of California.

Trivia pacifica, *Gray.*
Gulf of California.

Trivia sanguinea, *Gray.*
Gulf of California.

Trivia Solandri, *Gray.*
Gulf of California.

Strombus gracilior, *Sby.*
Gulf of California.

Strombus granulatus, *Swains.*
Gulf of California.

Conus princeps, *Linn.*
Gulf of California.

Conus gladiator, *Brod.*
Panama.

Conus puncticulatus, *Hwass.*
Gulf of California.

Solarium granulatum, *Lam.*
Gulf of California.

Natica unifasciata, *Lam.*
Panama.

Natica maroccana, *Chemn.*
Panama.

Natica uber, *Val.*
Gulf of California.

Natica catenata.
Gulf of California.

Ranella cælata, *Brod.*
Panama.

Ranella nitida, *Brod.*
Panama.

Persicula phrygia, *Cpr.*
Gulf of California.

Oliva angulata, *Lam.*
Acapulco.

Oliva subangulata.
Gulf of California.

Olivella dama, *Mawe.*
Gulf of California.

Olivella anazora, *Ducl.*
Gulf of California.

Olivella gracilis, *Gray.*
Gulf of California.

Olivella intorta, *Cpr.*
Bodega Bay, California.

Olivella bœtica, *Cpr.*
California.

Olivella semistriata, *Gray.*
Panama.

Olivella tergina, *Ducl.*
Gulf of California.

Olivella undatella, *Lam.*
Gulf of California.

Olivella volutella, *Lam.*
Panama.

Purpura biserialis, *Blainv.*
Panama.

Purpura melo, *Ducl.*
Panama.

Purpura triangularis, *Blainv.*
Gulf of California.

Cuma kiosquiformis, *Ducl.*
Panama.

Monoceros brevidentatum, *Gray.*
Panama.

Engina carbonaria, *Rev.*
Panama.

Engina Reeviana, *C.B.Ad.*
Panama.

Nitidella cribraria, *Lam.*
Nicaragua.

Columbella festiva, *Kien.*
Gulf of California.

Columbella fuscata, *Sby.*
Gulf of California.

Columbella hæmastoma, *Sby.*
Gulf of California.

Columbella major, *Sby.*
Gulf of California.

Nassa luteostoma, *B. & Sby.*
Panama.

Nassa Panamensis. *C. B. Ad.*
Panama.

Nassa scabriuscula, *Powis.*
Panama.

Nassa tegula, *Rve.*
Panama—Gulf of California.

Nassa versicolor, *C. B. Ad.*
Panama.

Anachis coronata, *Sby.*
Gulf of California.

Anachis costellata, *B. & Sby.*
Panama.

Anachis fluctuata, *Sby.*
Panama.

Anachis nigricans, *Sby.*
Panama.

Anachis pygmæa, *Sby.*
Panama—Gulf of California.

Anachis rugosa, *Sby.*
Panama.

Anachis varia, *Sby.*
Panama.

Columbella baccata, *Gask.*
Gulf of California.

Strombina gibberula, *Sby.*
Gulf of California.

Strombina maculosa, *Sby.*
Gulf of California,

Strombina solidula.
Gulf of California.

Pisania insignis, *Rve.*
Panama.

Pisania lugubris, *C. B. Ad.*
Panama.

Pisania ringens, *Rve.*
Panama.

Pisania sanguinolenta, *Ducl.*
Gulf of California.

Clavella distorta, *Bligh.*
Panama.

Murex plicatus, *Sby.*
Gulf of California.

Muricidea dubia, *Swains.*
Gulf of California.

Heterodonax bimaculatus.
Gulf of California.

Levenia coarctata.
Gulf of California.

Conella cedo-nulli.
Gulf of California,

Siphonalia pallida.
Gulf of California.

Physa aurantius, *Cpr.*
Near Acapulco, Mexico.

Patella viridula.
Valparaiso.

Patella Auracania. *D'Orb.*
Valparaiso.

Patella Auracania, *D'Orb. var.*
Valparaiso.

Monoceros citrinum. *Sby.*
Chili.

Littorina zebra.
Valparaiso.

Turbo niger. *Lesson.*
Chili.

Chlorostoma ater. *Lesson.*
Chili.

Ampullaria canalis. *D'Orb.*
Bolivia.

Collected by Robert E. C. Stearns.
Ischnoradsia regularis, *Cpr.*
Monterey, California.

Collected by Robert E. C. Stearns.
Chaetopleura Hartwegii, *Cpr.*
Monterey, California.

Collected by Robert E. C. Stearns.
Leptochiton rugatus, *Cpr.*
Monterey, California.

Collected by Robert E. C. Stearns.
Ischnochiton radians, *Cpr.*
Monterey, California.

Collected by Robert E. C. Stearns.
Trachydermon fallax, *Cpr.*
Monterey, California.

Collected by Robert E. C. Stearns.
Lepidopleurus Nuttalli, *Cpr.*
Monterey, California.

Collected by Robert E. C. Stearns.
Nacella vernalis, *Dall. (Mss.)*
Monterey, California.

From Robert E. C. Stearns.
Pisidium occidentale, *Newc.*
Mendocino County, California.

From Robert E. C. Stearns.
Pisidium Harfordiana, *Prime,*
Mendocino County, California. [*Mss.*

Collected by Robert E. C. Stearns.
Katherina tunicata, *Wood.*
Lobitas, California.

Collected by Robert E. C. Stearns.
Tonicia lineata, *Wood.*
Monterey, California.

Collected by Robert E. C. Stearns.
Mopalia muscosa, *Gld.*
Baulines Bay, California.

Collected by Robert E. C. Stearns.
Mopalia Wossnessenskii, *Midd.*
Monterey, California.

Collected by Robert E. C. Stearns.
Mopalia Hindsii, *Gray.*
San Francisco Bay, California.

Collected by Robert E. C. Stearns.
Mopalia Merckii, *Midd.*
Monterey, California.

Collected by Robert E. C. Stearns.
Nuttallina scabra, *Ree.*
Purissima, California.

Collected by Robert E. C. Stearns.
Lepidoradsia Magdalensis, *Hds.*
Monterey, California.

From Robert E. C. Stearns.
Bulimus albus, *Sby.*
Chili.

From Robert E. C. Stearns.
Bulimus Bridgesi, *Pfr.*
Huasco, Chili.

From Robert E. C. Stearns.
Bulimus Broderipi, *Sby.*
Chili.

From Robert E. C. Stearns.
Bulimus Coquimbensis, *Brod.*
Chili.

From Robert E. C. Stearns.
Bulimus corneus, *Sby.*
Nicaragua.

From Robert E. C. Stearns.
Bulimus coturnix, *Sby.*
Chili.

From Robert E. C. Stearns.
Bulimus crenulatus, *Pfr.*
Chili.

From Robert E. C. Stearns.
Bulimus elegans, *Pfr.*
Chili.

From Robert E. C. Stearns.
Bulimus membranaceus, *Phil.*
Nicaragua.

From Robert E. C. Stearns.
Bulimus pachychilus, *Pfr.*
Chili.

From Robert E. C. Stearns.
Bulimus pruinosus, *Sby.*
Chili.

From Robert E. C. Stearns.
Bulimus pupiformis. *Brod.*
Chili.

From Robert E. C. Stearns.
Bulimus reflexus, *Pfr.*
Chili.

From Robert E. C. Stearns.
Bulimus rosaceus, *King.*
Chili.

From Robert E. C. Stearns.
Bulimus rhodacme, *Pfr.*
Chili.

From Robert E. C. Stearns.
Bulimus erythrostoma. *Sby.*
Chili.

From Robert E. C. Stearns.
Bulimus variegatus, *Pfr.*
Chili.

From Robert E. C. Stearns.
Bulimus undatus. *Fer.*
Nicaragua.

From Robert E. C. Stearns.
Bulimus undatus, *var.*
Nicaragua.

From Robert E. C. Stearns.
Bulimus undatus.
Near Altata, Mexico.

Collected by Robert E. C. Stearns.
Terebratella caurina, *Gld.*
Near San Francisco, California.

Collected by Robert E. C. Stearns.
Pholadidea penita, *Conr.*
Monterey, California.

Collected by Robert E. C. Stearns.
Pholadidea ovoidea, *Gld.*
Monterey, California.

Collected by Robert E. C. Stearns.
Netastomella Darwinii, *Sby.*
Monterey, California.

Collected by Robert E. C. Stearns.
Parapholas Californica, *Conr.*
Monterey, California.

Collected by Robert E. C. Stearns.
Saxicava pholadis, *Linn.*
Monterey, California.

Collected by Robert E. C. Stearns.
Platyodon cancellatum, *Conr.*
Half Moon Bay, California.

Collected by Robert E. C. Stearns.
Cryptomya Californica, *Conr.*
San Francisco Bay, California.

Collected by Robert E. C. Stearns.
Lyonsia Californica, *Conr.*
Purissima, California.

Collected by Robert E. C. Stearns.
Machaera patula, *Dixon.*
Santa Cruz, California.

Collected by Robert E. C. Stearns.
Macoma secta, *Conr.*
Bodega Bay, California.

Collected by Robert E. C. Stearns.
Macoma nasuta, *Conr.*
Bodega Bay, California.

Collected by Robert E. C. Stearns.
Macoma inconspicua, *Br. & Sby.*
San Francisco Bay, California.

Collected by Robert E. C. Stearns.
Cumingia Californica, *Conr.*
Monterey, California.

Collected by Robert E. C. Stearns.
Donax Californicus, *Conr.*
San Diego, California.

Collected by Robert E. C. Stearns.
Amiantis callosa, *Conr.*
San Diego, California.

Collected by Robert E. C. Stearns.
Pachydesma crassatelloides,
San Diego, California.   [*Conr.*

Collected by Robert E. C. Stearns.
Psephis tantilla, *Gld.*
Monterey, California.

Collected by Robert E. C. Stearns.
Chione simillima, *Sby.*
San Diego, California.

Collected by Robert E. C. Stearns.
Chione fluctifraga, *Sby.*
San Diego, California.

Collected by Robert E. C. Stearns.
Tapes staminea, *Conr.*
California.

Collected by Robert E. C. Stearns.
Tapes staminea, *var.* Petitii,
California.   [*Desh.*

Collected by Robert E. C. Stearns,
Tapes staminea, *var.* ruderata,
California.   [*Desh.*

Collected by Robert E. C. Stearns.
Tapes staminea, *var.* diversa,
Tomales Bay, California.   [*Sby.*

Collected by Robert E. C. Stearns.
Saxidomus Nuttalli, *Conr.*
Vancouver Island.

Collected by Robert E. C. Stearns.
Rupellaria lamellifera, *Conr.*
Monterey, California.

Collected by Robert E. C. Stearns.
Petricola carditoides, *Conr.*
Monterey, California.

Collected by Robert E. C. Stearns.
Chama exogyra, *Conr.*
San Diego, California.

Collected by Robert E. C. Stearns.
Chama pellucida, *Sby.*
Monterey, California.

Collected by Robert E. C. Stearns.
Cardium corbis, *Martyn.*
California.

Collected by Robert E. C. Stearns.
Lazaria subquadrata, *Cpr.*
Monterey, California.

Collected by Robert E. C. Stearns.
Lucina Californica, *Conr.*
Monterey, California.

Collected by Robert E. C. Stearns.
Diplodonta orbella, *Gld.*
Monterey, California.

Collected by Robert E. C. Stearns.
Kellia Laperousii, *Desh.*
Monterey, California.

Collected by Robert E. C. Stearns.
Mytilus Californianus, *Conr.*
California.

Collected by Robert E. C. Stearns.
Septifer bifurcatus, *Rec.*
Monterey, California.

Collected by Robert E. C. Stearns.
Modiola fornicata, *Cpr.*
Monterey, California.

Collected by Robert E. C. Stearns.
Adula falcata, *Gld.*
Monterey, California.

Collected by Robert E. C. Stearns.
Adula stylina, *Cpr.*
Lobitas, California.

Collected by Robert E. C. Stearns.
Pecten latiauritus, *Conr.*
Monterey, California.

Collected by Robert E. C. Stearns.
Hinnites giganteus, *Gray.*
Baulines Bay, California.

Collected by Robert E. C. Stearns.
Bulla nebulosa, *Gld.*
San Diego, California.

Collected by Robert E. C. Stearns.
Tornatella punctocælata, *Cpr.*
Monterey, California.

From Robert E. C. Stearns.
Helix Vancouverensis, *Lea.*
Sitka.

Collected by Robert E. C. Stearns.
Helix Dupetithouarsi, *Desh.*
Monterey, California.

Collected by Robert E. C. Stearns.
Helix Californiensis, *Lea.*
Monterey, California.

Collected by Robert E. C. Stearns.
Helix arrosa, *Gld.*
San Mateo County, California.

Collected by Robert E. C. Stearns.
Helix Columbiana, *Lea.*
California — Sitka.

Collected by Robert E. C. Stearns.
Pupilla Californica, *Rowell.*
Monterey, California.

Collected by Robert E. C. Stearns.
Physa diaphana, *Tryon.*
Alameda County, California.

From Robert E. C. Stearns.
Dentalium Indianorum, *Cpr.*
Vancouver Island, B. C.

Collected by Robert E. C. Stearns.
Chiton tunicata, *Wood.*
Purissima, California.

Collected by Robert E. C. Stearns.
Chiton lineata, *Wood.*
Monterey, California.

Collected by Robert E. C. Stearns.
Chiton muscosa, *Gld.*
Baulines Bay, California.

Collected by Robert E. C. Stearns.
Chiton Hindsii, *Gray.*
San Francisco Bay, California.

Collected by Robert E. C. Stearns.
Chiton scabra, *Rve.*
Monterey, California.

Collected by Robert E. C. Stearns.
Chiton Magdalensis, *Hds.*
Monterey, California.

Collected by Robert E. C. Stearns.
Chiton regularis, *Cpr.*
Monterey, California.

Collected by Robert E. C. Stearns.
Nacella incessa, *Hds.*
Baulines Bay, California.

Collected by Robert E. C. Stearns.
Nacella depicta, *Hds.*
Monterey, California.

Collected by Robert E. C. Stearns.
Acmæa patina, *Esch.*
Purissima, California.

Collected by Robert E. C. Stearns.
Acmæa pelta, *Esch.*
Purissima, California.

Collected by Robert E. C. Stearns.
Acmæa Asmi, *Midd.*
Baulines Bay, California.

Collected by Robert E. C. Stearns.
Acmæa persona, *Esch.*
Monterey, California.

Collected by Robert E. C. Stearns.
Acmæa scabra, *Nutt.*
Monterey, California.

Collected by Robert E. C. Stearns.
Acmæa spectrum, *Nutt.*
Purissima, California.

Collected by Robert E. C. Stearns.
Lottia gigantea, *Gray (Sby?)*
Monterey, California.

Collected by Robert E. C. Stearns.
Scurria mitra, *Esch.*
Monterey, California.

Collected by Robert E. C. Stearns.
Rowella radiata, *Cp.*
Monterey, California.

Collected by Robert E. C. Stearns.
Fissurella volcano, *Rve.*
Monterey, California.

Collected by Robert E. C. Stearns.
Glyphis aspera, *Esch.*
Monterey, California.

Collected by Robert E. C. Stearns.
Haliotis Cracherodii, *Leach.*
Monterey, California.

Collected by Robert E. C. Stearns.
Haliotis rufescens, *Sw.*
Monterey, California.

Collected by Robert E. C. Stearns.
Phasianella pulloides, *Cpr.*
Monterey, California.

From Robert E. C. Stearns.
Pomaulax undosus, *Wood.*
San Diego, California.

Collected by Robert E. C. Stearns.
Pachypoma gibberosum, *Chem.*
Monterey, California.

Collected by Robert E. C. Stearns.
Leptothyra sanguinea, *Cpr.*
Monterey, California.

Collected by Robert E. C. Stearns.
Leptothyra bacula, *Cpr.*
Monterey, California.

Collected by Robert E. C. Stearns.
Chlorostoma funebrale, *A. Ad.*
Baulines Bay, California.

Collected by Robert E. C. Stearns.
Chlorostoma brunneum, *Phil.*
Monterey, California.

From Robert E. C. Stearns.
Chlorostoma aureotinctum, *Fbs.*
San Diego, California.

From Robert E. C. Stearns.
Omphalius fuscescens, *Phil.*
San Diego, California.

Collected by Robert E. C. Stearns.
Calliostoma canaliculatum, *Mart.*
Monterey, California.

Collected by Robert E. C. Stearns.
Calliostoma costatum, *Mart.*
Monterey, California.

Collected by Robert E. C. Stearns.
Calliostoma annulatum, *Mart.*
Monterey, California.

Collected by Robert E. C. Stearns.
Phorcus pulligo, *Mart.*
Monterey, California.

Collected by Robert E. C. Stearns.
Gibbula succincta, *Cpr.*
Monterey, California.

Collected by Robert E. C. Stearns.
Margarita pupilla, *Gld.*
Monterey, California.

Collected by Robert E. C. Stearns.
Crepidula adunca, *Sby.*
Monterey, California.

Collected by Robert E. C. Stearns.
Crepidula navicelloides, *Nutt.*
Monterey, California.

Collected by Robert E. C. Stearns.
Crepidula navicelloides, *var.* [nummaria, *Gld.*
Monterey, Cal.

Collected by Robert E. C. Stearns.
Crepidula navicelloides, *var.* fim-[briata, *Rve.*
Monterey, Cal.

Collected by Robert E. C. Stearns.
Crepidula navicelloides, *var.* ex-[planata, *Gld.*
Monterey, Cal.

Collected by Robert E. C. Stearns.
Cerithidea sacrata, *Gld.*
California.

Collected by Robert E. C. Stearns.
Bittium filosum, *Gld.*
Monterey, California.

Collected by Robert E. C. Stearns.
Litorina planaxis, *Nutt.*
Bodega, California.

From Robert E. C. Stearns.
Litorina Sitkana, *Phil.*
Vancouver Island, Sitka.

Collected by Robert E. C. Stearns.
Litorina scutulata, *Gld.*
Purissima, California.

Collected by Robert E. C. Stearns.
Bythinella Binneyi, *Tryon.*
Bauliaes Bay, California.

Collected by Robert E. C. Stearns.
Diala marmorea, *Cpr.*
Monterey, California.

Collected by Robert E. C. Stearns.
Trivia Californiana, *Gray.*
Monterey, California.

Collected by Robert E. C. Stearns.
Erato vitellina. *Hds.*
Monterey, California.

Collected by Robert E. C. Stearns.
Mitromorpha filosa, *Cpr.*
Monterey, California.

From Robert E. C. Stearns.
Conus Californicus, *Hinds.*
San Diego, California.

Collected by Robert E. C. Stearns.
Odostomia gravida, *Gld.*
Monterey, California.

From Robert E. C. Stearns.
Prieue Oregonensis, *Redf.*
Vancouver Island, B. C.

Collected by Robert E. C. Stearns.
Marginella Jewettii, *Cpr.*
Monterey, California.

Collected by Robert E. C. Stearns.
Marginella subtrigona, *Cpr.*
Monterey, California.

Collected by Robert E. C. Stearns.
Marginella regularis, *Cpr.*
Monterey, California.

Collected by Robert E. C. Stearns.
Marginella pyriformis, *Cpr.*
Monterey, California.

Collected by Robert E. C. Stearns.
Olivella biplicata, *Sby.*
San Diego, California.

Collected by Robert E. C. Stearns.
Nassa mendica. *Gld.*
Monterey, California.

From Robert E. C. Stearns.
Amycla carinata. *Hds.*
San Diego, California.

Collected by Robert E. C. Stearns.
Amycla carinata, *var?* **Hindsii,**
California.   [*Rve.*

Collected by Robert E. C. Stearns.
Amycla gausapata, *Gld.*
California.

Collected by Robert E. C. Stearns.
Amycla tuberosa, *Cpr.*
Monterey, California.

Collected by Robert E. C. Stearns.
Amphissa corrugata. *Rve.*
Monterey, California.

From Robert E. C. Stearns.
Purpura crispata, *Chem.*
Vancouver Island, Sitka.

From Robert E. C. Stearns.
Purpura canaliculata, *Duel.*
Vancouver Island, Sitka.

From Robert E. C. Stearns.
Purpura saxicola, *Val.*
Vancouver Island, B. C.

Collected by Robert E. C. Stearns.
Purpura saxicola, *var.* **ostrina.**
Bodega Bay, California. [*Gld.*

Collected by Robert E. C. Stearns.
Purpura saxicola, *var?* **emargin-**
Monterey, Cal. [ata, *Desh.*

From Robert E. C. Stearns.
Purpura trisérialis, *Blainv.*
San Diego, California.

Collected by Robert E. C. Stearns.
Monoceros engonatum, *Conr.*
Baulines Bay, California.

Collected by Robert E. C. Stearns.
Monoceros lapilloides, *Conr.*
Monterey, California.

Collected by Robert E. C. Stearns.
Ocinebra lurida, *Midd.*
California.

Collected by Robert E. C. Stearns.
Ocinebra lurida, *var.* **aspera,**
California. [*Baird.*

Collected by Robert E. C. Stearns.
Ocinebra lurida, *var.* **munda,** *Cpr.*
California.

Collected by Robert E. C. Stearns.
Ocinebra interfossa, *Cpr.*
California.

Collected by Robert E. C. Stearns.
Ocinebra interfossa, *var.* **atropur-**
California. [purea, *Cpr.*

Collected by Robert E. C. Stearns.
Cerostoma foliatum, *Gmel.*
California.

From Robert E. C. Stearns.
Cerostoma Nuttalii, *Conr.*
San Diego, California.

From Robert E. C. Stearns.
Chrysodomus dirus, *Rve.*
California — Vancouver Island.

From Robert E. C. Stearns.
Volutharpa ampullacea, *Midd.*
Aleutian Islands.

Collected by Robert E. C. Stearns.
Purpura canaliculata, *Duclos.*
Alaska.

www.ingramcontent.com/pod-product-compliance
Lightning Source LLC
Chambersburg PA
CBHW031825230426
43669CB00009B/1226